Swimming Upstream

Teaching and Learning Psychotherapy in a Biological Era

Other Books by Jerry M. Lewis, M.D.

No Single Thread: Psychological Health in Family Systems, 1976
(with W. R. Beavers, J. T. Gossett, & V. A. Phillips)

To Be a Therapist: The Teaching and Learning, 1978

How's Your Family?, 1979

Psychiatry in General Medical Practice, 1979
(edited with G. Usdin)

The Family: Evaluation and Treatment, 1980
(edited with C. K. Hofling)

Treatment Planning in Psychiatry, 1982
(edited with G. Usdin)

To Find a Way: The Outcome of Hospital Treatment of Disturbed Adolescents, 1983 *(with J. T. Gossett & F. D. Barnhart)*

The Long Struggle: Well-functioning, Working-class Black Families, 1983 *(with J. G. Looney)*

The Birth of the Family: An Empirical Inquiry, 1989

Swimming Upstream

Teaching and Learning Psychotherapy in a Biological Era

Jerry M. Lewis, M.D.

Brunner/Mazel, Publishers • New York

Library of Congress Cataloging-in-Publication Data

Lewis, Jerry M.

Swimming upstream : teaching and learning psychotherapy in a biological era / Jerry M. Lewis.

p. cm.

Includes bibliographical references.

Includes index.

ISBN 0-87630-612-1

1. Psychotherapy—Study and teaching. 2. Psychotherapy. I. Title.

[DNLM: 1. Psychotherapy—education. 2. Teaching—methods. WM 18 L674s]

RC459.L49 1991

616.89'14'0711—dc20

DNLM/DLC

for Library of Congress

90-2671

CIP

Published by

BRUNNER/MAZEL, INC.

19 Union Square West

New York, New York 10003

Design by Thomas A. Bérubé

Manufactured in the United States of America

10 9 8 7 6 5 4 3 2 1

Finding love in marriage and pleasure in work fills the cup. If, in addition, one has truly good friends throughout the years, the cup overflows. I would like, therefore, to dedicate this book to a circle of friends who, each in his or her own way, sustain me.

Doyle and Sarah Carson
E. John Caruso
John Gossett
Tom and Lynn Lombardo
Peter and Sharon Martin
Gene and Cecile Usdin
Bob and Jean Van de Wetering
Harold and Gladys Visotsky
Michael and Ruth Zales

Contents

Preface

A distinguished colleague, upon reading this manuscript, made two suggestions. The first was that I offer the reader an explanation for what he considered to be the unusual amount of self-disclosure in the text. The second was that I emphasize even more strongly my awareness that the urgency that I attach to the continuation (and growth) of systematic teaching of psychotherapeutic skills is based primarily on my belief that those skills are central to the core identity of the psychiatrist. On the rare occasions when this friend has given advice, he has demonstrated an unerring accuracy. Therefore, I take the opportunity provided by an author's preface to follow his suggestions.

The experiential base from which this book is written is my seminar for psychiatric residents which emphasizes that self-disclosure with colleagues is an important aspect of becoming a therapist. The ability to look at and listen to one's own work along with one's peers is important in the maturation process. In order to construct a context in which it is possible to learn from each other, I share many of my own psychotherapeutic experiences. More than this, however, I share personal experiences when they seem appropriate to the teaching-learning process.

In writing this book, I have replicated this process. It has seemed to me that the structure of this book should parallel the structure of the seminar in which young therapists can become comfortable with being open about what they actually say and do in psychotherapy sessions. For those who would approach the task of teaching psychotherapy skills with less disclosure of self, I say only that this has been my way, and I share it here as just that—not the only way.

As to the second suggestion, I wish to emphasize my belief that the characteristics that distinguish psychiatry from other disciplines include an emphasis on the role of psychosocial factors in psychopathology and the use of psychotherapeutic interventions.

My reading of the psychotherapy outcome literature leads me to the conclusion that psychotherapy is clearly helpful in a variety of syndromes. More recently, research reports even suggest what type of psychotherapy may be most helpful in specific syndromes.

Research that illuminates the effects of various psychotherapy training programs is lacking, however. The systematic, experiential approaches to the training of beginning psychotherapists described in this text are not supported by empirical studies, but rest on my belief, my experience, and my commitment to experiential training. The readers must conclude for themselves whether the seminar I describe is a process worthy of their commitment—particularly at a time when the future of psychotherapy is in some peril.

Jerry M. Lewis

Acknowledgments

Writing a book, even one that is very personal, occurs in a context that either facilitates or impedes. One of the major advantages I have had throughout my career has been the presence of facilitating contexts. I have acknowledged on the occasion of other books my immense gratitude to my wife, Pat, for her central role in providing a context that supports my work generally and my writing specifically.

On other occasions I have also tried to make clear the debt I owe my colleagues at Timberlawn for the professional context they provide. At the Hospital, Doyle Carson, Keith Johansen, Byron Howard, and Mark Blotcky provide leadership to a system that facilitates. At the Foundation, John Gossett and Margaret Tresch Owen lend friendship, respect, and support that allows me to continue my work.

At a somewhat more operational level, Virginia Austin Phillips offers me superb editorial assistance, John Gossett provides a different level of critical overview, and Nannette Bruchey prepares the manuscripts. For all of this caretaking I am truly grateful.

Swimming Upstream

Teaching and Learning
Psychotherapy in a
Biological Era

CHAPTER 1

Psychotherapy: An Uncertain Future

This book reflects 25 years of experience teaching basic psychotherapy skills to beginning residents and other professionals. An earlier book was published over 10 years ago (Lewis, 1978), and since that time I have learned more about teaching, the seminar has broadened to include additional components of the psychotherapeutic process, and it seems appropriate to summarize the current state of my experience.

The major factor in the decision to write this book, however, is my deep concern about what may happen—or is happening already—to the role of learning psychotherapy in the education of psychiatrists. Case conferences in some academic centers have senior residents who respond to the request for a clinical formulation and treatment plan with only a DSM-III-R diagnosis and a discussion of various psychopharmacologic approaches to the patient. Professional debates argue whether or not psychodynamics and psychotherapy should be a part of a residency program's curriculum. Possibly the next generation of psychiatrists is being trained to know even less than we did about the process of psychotherapy.

This development is alarming; the reasons are diverse and powerful. In this chapter I wish to explore briefly some of the more obvious reasons, to emphasize the various effects on psychiatric practice brought about by the economic crisis in health care, to examine the impact of the rapid growth in the neurosciences, and to explore the difficulties involved in using a complex model of psychopathology such as a biopsychosocial approach. Finally, I will speak to the problems that arise in maintaining a balance between our own collective definition of psychiatry's core distinctions and responding to forces from without that seek to shape what we do—thus, how we are defined. The issues for me are what are psychiatry's core distinctions and how can we avoid having them molded entirely by forces impinging upon us from without.

Even a casual reader of newspapers and periodicals knows of the growing wave of resentment of rising health care costs. Each day brings yet another statement that this country must change its approach to financing health care; that the 11+ percent of the GNP we spend is far more than other industrialized nations; that over 30 million Americans are without any form of health insurance; that only the wealthy receive the full advantage of this country's preeminence in scientific medicine. The suggestion that we adopt a British type national health insurance financed almost entirely by general tax revenues is frequently counterbalanced by analyses of the British system with its long lists of those waiting for elective surgery, decaying hospital physical plants, and the less than intensive treatment, for example, of some severe congenital defects in children.

At the time of this writing, it is even more popular to consider the adoption of the Canadian system, but here, too, appear analyses of the deficiencies of that system. Recent surveys indicate, however, that Americans are less satisfied with their health care system than are either the British or the Canadians (Freudenheim, 1989; Whitney, 1989). Clearly, the time is ripe for changes, but nobody really knows what will work.

We know that we can't do everything we know how to do and meet other government objectives. Years ago, the medical economist Fuchs (1974) entitled his essays on this topic, *Who Shall Live?* Ex-Governor Lamm (1989) of Colorado speaks eloquently about the need to define "appropriate" health care and construct a system that insures that all Americans receive it. His persuasive voice is but one among many, but the difficulty is in the definition of "appropriate." Such a definition will surely mean that some of what we do will be defined as "less appropriate" or even "inappropriate."

One experiment in defining appropriate health care is now under way in the Oregon medical rationing system for the indigent. Although all the details are not available, early press reports indicate that when the money is limited the treatment of certain diseases will be financed, others will not be. The ethical issues are profound, and it has been said that it is like dropping bombs on people you cannot see and do not know (Gross, 1989).

Others write about our failure to approach an even more basic issue: the need to decide what ends or purposes we believe a health care system should serve. Churchill (1989), for example, emphasizes the profound ambivalence underlying this unwillingness to decide, perhaps because most Americans oppose additional taxes to enact that right.

It is also clear that we have tremendous expectations regarding health care; the advances in scientific medicine are well publicized because they have great impact upon our collective fear of dying before our time.

The problems are complex, and the attempted solutions will almost certainly change the basic fabric of medicine as we know it. Although much more is being said about these issues by individuals more knowledgeable than this writer, the focus for purposes of this chapter is on how such changes will influence psychiatry and, more specifically, the practice of psychotherapy.

We do not know. Although the American Psychiatric Association is vigorously leading the attempt to preserve and improve the position of psychiatric patients and has been joined in this battle by increasingly powerful advocacy groups, we simply do not know what the outcome will be. There are, however, some ominous signs. One is the poor faring of psychiatry in the proposed Resource-Based Relative Value Scale (RBRVS). Although the intent of this effort was to increase payment for cognitive services (internists, family medicine specialists, etc.) and decrease payment for procedure-oriented specialists, psychiatry, one of the most cognitively oriented specialties, will be hurt even more than some branches of surgery (Evelyn, 1989). The American Psychiatric Association has taken the position that the evaluation of practice costs was inadequate and that more difficult clinical situations were not covered in the vignettes used to assess the time and skill required of the psychiatrist. Despite this effort, there is doubt that any really basic change in the RBRVS will be made.

What does this mean? First, the proposed use of the scale is for Medicare payments, a system that has characteristically given psychiatric treatment very low funding. The concern, however, is that the RBRVS, if adopted by Medicare as proposed, will become the standard for all third-party payers. The issue is not the reimbursement of psychiatrists relative to other medical specialists; rather, it is the continuing operation of the stigma regarding mental illness and the continuing low assessment of psychiatry within medicine generally. The joke about "real" doctors and psychiatrists is the ambiance in which relative values are assigned.

If the proposed RBRVS is not exactly a rosy sign for psychiatry's future, what about insurance coverage for both outpatient and inpatient psychiatric care? I have seen no systematic studies and, thus, must rely on personal experience: Policies that support psychiatric hospitalization do so only for very short hospital stays. Most often the provisions of such policies allow the pharmacologic management of acute clinical episodes like a major depression with significant risk of suicide or a serious manic episode. They fail miserably, however, to provide for appropriate treatment of many more chronic conditions such as severe borderline states, psychoses which do not respond to psychotropic agents, and severely disturbed patients for whom the establishment of a treatment alliance takes many months.

Recently, for example, I saw in consultation a hospitalized 16-year-old girl with a two-year history of gradual change involving school refusal, intravenous drug use, prostitution, and participation in a variety of criminal activities. She vigorously denied any need for treatment, blamed her parents and their unreasonable expectations for her situation, and had successfully resisted outpatient psychotherapy, a trial of antidepressives, family therapy, and several short-term hospitalizations.

I thought she needed intensive hospitalization until an alliance with her and her family could be forged, a treatment relationship with a psychotherapist established, and appropriate psychotropic agents tried. Following this initial stage of treatment, it was likely that she would require a series of increasingly less structured residential treatment facilities, all the while maintaining continuity of care with both an individual and a family therapist.

This youngster may represent the exception, the child who cannot be treated without a very inclusive long-range approach. The irony is that up until a few years ago the very agency for which both her parents work provided employees with insurance coverage that underwrote such treatment. Because of economic considerations, the coverage for psychiatric disturbances was reduced to 30 days a year of hospitalization and 30 outpatient visits each year. Psychiatric disturbances came to be excluded from coverage under the major medical illness portion of their policy. Thus, one is left with a life that is bleeding out and without the provisions to help.

Existing research data are sufficient to document the most probable untreated outcome for this youngster (Robins, 1966; Rutter & Quinton, 1984). Probably, she will survive for some time, perhaps have a child whom she cannot mother adequately, and quite likely a transgenerational cycle will begin. Imprisonment or premature death through murder, suicide, or disease is a real possibility. Our own follow-up study suggests that the lives of over three-fourths such seriously disturbed adolescents can be salvaged with appropriate treatment (Gossett, Lewis, & Barnhart, 1983, p. 68).

This clinical dilemma is faced every day by clinicians harried by third-party payers pressing for a rapid discharge. Although such pressure may facilitate creative and workable approaches to the treatment of some, for others it means simply no real treatment is possible.

The same movement to shorten and circumscribe psychiatric treatment is present regarding outpatient services also. Experience with traditional health insurance affords little optimism about how psyschiatric treatment will fare in the future, particularly if it involves intensive outpatient psychotherapy or hospitalization other than the briefest sort.

The outlook is no different regarding managed health care. Here, also, no systematic studies document trends, but my experience with these intermediaries suggests that the contracts they establish provide only relatively short-term outpatient psychotherapy, at a level of once-per-week for a total of 15-30 sessions a year, and only very brief crisis-oriented inpatient care. The chronic, severely dysfunctional patient often requires intensive treatment, but is deprived of adequate psychiatric care.

Finally, what is the outlook for psychiatric patients when there is explicit rationing of health care for the poor? At the time of this writing, Oregon is the only state that ranks all illnesses in terms of which services will be underwritten. Newspaper reports indicate that prenatal care and the immunization of children will be high priorities when such rankings are made. Eating disorders, most often treated by psychiatrists, will be a low priority, although, in general, mental disorders and chemical dependency are covered by other state agencies. How will pervasive developmental defects of childhood be ranked? What consideration will be given to the recent Institute of Medicine report (Johnson, 1989) indicating that at least 7.5 million—12 percent of the nation's population under age 18—suffer from a mental disorder or emotional disturbance requiring treatment, with only one of three such children receiving treatment in 1985. Mental health professionals are deeply concerned that mental health service may be considered separate from other services and, and if included in any forced ranking, would not be accorded the funding so badly needed.

Although the final funding of psychiatric treatment is yet to be determined, the information at hand is not encouraging. Whether it is the Resource-Based Relative Value Scale, traditional health insurance, explicit rationing, or managed health care programs, pressure mounts to treat patients briefly in an acute illness or episodic model. Patients whose illnesses are chronic and require long and intensive treatment will receive grossly inadequate treatment. The economic forces pushing toward brief treatment will mold treatment in the direction of relying only on psychotropic agents. Psychotherapy, particularly if it is required more than once a week for any extended period of time, will not be covered.

If this scenario is valid, organized psychiatry must continue to do everything possible to avoid this external definition of how psychiatrists treat patients, and ultimately, conceptualize their disorders. In addition to these political activities, psychiatry needs to emphasize internally and through the curricula of our residency training programs, the broadly based and comprehensive models of psychopathology and the central role of psychotherapy as core distinctions. To allow the learning of psychotherapy to become an auxiliary elective, to be studied only by those who are so inclined, participates, however inadvertently, in a definition of our profession dictated by outside forces.

And such oversight is easy in these heady days of remarkable advances in the neurosciences. The biological revolution of recent decades has added enormously to our knowledge of the central nervous system and directly influenced diagnosis and treatment approaches. The practice of psychiatry has been changed forever, and clinicians have an ever increasing number of diagnostic techniques and drugs with which to help patients. Advances in neurobiology, genetics, and immunology portend new insights for the future. Never before has psychiatry had such remarkable data with which to understand and treat mental illness.

Our challenge is to integrate the new biological findings with equally important psychological and social system findings. As Eisenberg (1988) has said, psychiatry is moving from brainlessness to mindlessness. Despite the evidence that psychological, family, and cultural factors influence each of the major mental illnesses, we often hear statements like, "Now that we know that schizophrenia is a brain disease. . . . "

This form of evangelical reductionism is not new. My generation of psychiatrists participated in several earlier periods in which the collective reductionistic fever ran high. These periods can be described as bandwagons: intensely exciting movements that dominate the entire field. They are characterized by a shared belief that definitive answers are now available, answers as to the "real" causes of psychiatric disturbances. Thus, all that remained was "a little more research," devising treatment interventions based upon the new and "definitive" facts, and training appropriate numbers of clinicians in their use.

Funding for research is heavily influenced by whatever system of variables rides the bandwagon. The curricula of residency programs come to be skewed in a similar way. Other systems of variables lose their importance; the excitement and ferment invite all associated with treatment to climb aboard.

During my residency, the bandwagon was psychoanalysis. The first cause of all psychiatric syndromes was psychodynamics. Depression, the schizophrenias, the neuroses—every syndrome was viewed as a direct reflection of psychodynamic conflict. Further, each syndrome had a specific psychodynamic conflict, and in the hands of the most convinced, if that conflict were not present, the diagnosis was in doubt. Departments of psychiatry reflected this reductionistic orientation; the contributions of other etiologic variables received scant attention. Great excitement overrode doubt. Clinicians who held to the earlier biologic orientation with its insulin coma and electroshock treatments were seen as clearly out of date. If enough psychodynamically oriented therapists could be trained, the schizophrenias would yield to accurate interpretations.

Something changed this. The causes of the change are difficult to know but, in part, psychodynamic psychotherapy failed to live up to such grandiose expectations. Only a few patients, rather than all the patients with more severe syndromes, were treated successfully. The country's political climate changed, and greater attention was given to deleterious social forces. The community mental health movement was born, and a new fever emerged: the social system bandwagon. First causes were thought to be found in destructive social dynamics: poverty, poor housing, nutrition, prejudice, and, in particular, family pathology. Family therapy came to be a central organizing perspective for many mental health professions, although never achieving first-rank status in psychiatry.

Within the family therapy movement, all psychiatric syndromes were understood as disturbances of context. Psychiatric symptoms were thought to reflect dysfunctional family systems and the heroic attempts by the symptomatic family member to prevent total disintegration of the family. Therapists who made causal explanations that emphasized the role of individual psychodynamics were thought to be misinformed. If only enough family therapists could be trained, the schizophrenias would yield to unlocking family double binds.

Again, something changed. However valuable this approach was in some instances, most major mental disorders did not yield to family therapy. Gradually, that bandwagon ground to a halt, and psychiatry climbed aboard a new bandwagon—the neurosciences. It may be too simple to say that the change was brought about by the rapid advances in neuroscience research. Many of the early psychotropic agents preceded the explosion of research. DSM-III, with its clear biologic orientation, is as much a result as it is a cause of the resurgency of the biologic movement. Although the impact of large sociocultural changes on psychiatric doctrine is difficult to ascertain, biology's inherent conservatism parallels, in a rough way, the concurrent conservative political climate of the country.

The oversimplification, reductionism, and linearity expressed by the most evangelical of current biological psychiatrists are the same cognitive characteristics demonstrated earlier by the most fervent priests of psychodynamics and social psychiatry. There would seem to be a lust for the certainty of simple explanations, a turning away from the complexity of mental illness.

But we face a recurring paradox. While each new explanation brings new perspectives to psychiatry, exciting data, and broadened understanding of behavior and mental illness, yet, instead of these new insights being integrated with the old insights, there is inevitable reductionism. Just as my training did not attend biologic variables, many current residents are being trained with little attention to developmental and social system variables. Departments of psychiatry are very likely to be led by neuroscientists; departmental research is primarily biologic; and often faculty role models are not experienced psychotherapists.

In most departments I am familiar with, the exciting advances in family system research and family therapy are scarcely noted. Few current residents are aware of the early findings of a recent Finnish study indicating that none of the adopted-away children of schizophrenic mothers developed severe psychopathology if adopted by a well-functioning family (Tienari et al., 1985). Marital and family therapies are done by social workers, and seldom does a resident achieve comfort or competence in understanding and intervening in these human systems.

Despite the real excitement of advances in the neurosciences, the development of highly sophisticated brain-measurement techniques, and the extraordinary effects of psychotropic agents, is it necessary to disregard what we already know? Is it necessary to reduce our commitment to a broadly based comprehensive psychiatry in which the learning of psychotherapeutic skills continues to be a prime goal of training the next generation of psychiatrists? The answer, of course, is no. It is not necessary, but it is already happening in some training programs. If the force of biologic reductionism is combined with that of the economic shaping of psychiatric treatment based on an acute illness model with brief, primarily biologic treatments, psychotherapy training may be in serious jeopardy.

An important aspect of our collective failure to integrate the useful data from previous insights with the exciting new biologic data is the difficulty involved in formulating and using complex models of causality. I teach our residents how to construct psychodynamic formulations, but I emphasize that the psychodynamic features must not stand alone and that a more broadly based clinical formulation involves serious attention to biological and social system variables. When we teach residents how to construct a broadly based formulation, we are teaching them which clinical data are important and how to think about those data. In that teaching effort we must be careful not to suggest simple formats of whatever kind.

For over a decade, Engel's (1980) biopsychosocial model has been the most attended complex model of causality. It is important to recall that Engel proposed this model not for psychiatry and psychopathology specifically, but as a replacement of the more reductionistic biomedical model of general medicine. Although the biopsychosocial model has received much acclaim, it has made very little impact on the practice of medicine or, more specifically, psychiatry. Why is this so? Undoubtedly, there are many factors, including the success of the older biomedical model in achieving new understanding of disease and remarkable breakthroughs in medical treatment. However, the biomedical model has been less successful in dealing with chronic disease wherein there is much to suggest that complex interactions of many variables operate. When the causal field becomes crowded with complex interactions, a more complex model of causality is needed, and systems theory rose to meet the needs of complex causality and is the foundation of Engel's biopsychosocial model.

It is hard to escape the conclusion that how we think about causality is a deeply ingrained characteristic. The simpler the causal explanation, the greater the sense of knowing, of certainty, and of a clearly defined path of action. Most of us relish that state of mind. The simpler models of causality free us for decisive actions. Two simple models of causality are everyday fare for psychiatrists: One is linear and the other circular. The linear model is illustrated, for example, in the concept that a husband's emotional distancing causes his wife to complain angrily at his lack of availability. The circular model states that the husband's distancing produces the wife's anger which, in turn, produces the husband's distancing. This circular conceptualization is that of a fairly simple system in which both participants' behaviors are simultaneously cause and effect. Little is gained by assigning the primary causal role to either participant; the interaction itself becomes the unit of causality.

Up to this point most clinicians feel comfortably at home. But when the interaction involves numerous parts, often at very different levels of complexity, the observer's sense of comfortable certainty departs. Beahrs' (1986) articulation and extension of the biopsychosocial model for psychiatry is very helpful and is illustrated in Figure 1.1. Variations of his illustrations of complex causality are noted in Figures 1.2, 1.3, and 1.4. Figure 1.1 represents the multiple interacting variables involved in complex causality. Figure 1.2 is Beahrs' original figure, and I suggest that the bold lines are the variables focused on by a biological psychiatrist. Figures 1.3 and 1.4 represent the different variables from the same field focused on by a psychodynamic psychiatrist and by a family psychiatrist.

Although even these figures simplify too much, they seem useful in addressing the increased complexity when one moves from linear and circular models to models that incorporate many variables. The figures also speak to Beahrs' central theme: We can never be truly certain in our understanding of a given patient's psychopathology because we are prisoners of our perspectives. If, however, it can be recognized that each perspective results in valuable understanding and that the perspectives are complementary rather than oppositional, we have moved into the real world of complex models.

Whatever our perspective, the very act of observing changes the interactions. A true and finite reality is unknowable because significant aspects of what we "see" are constructed by the act of seeing.

Thus, Beahrs emphasizes the approach Havens (1973) earlier called pluralism—the ability to use different models as each may fit best a given clinical situation. Truly integrating the biological, psychological, and social variables is beyond our capacity at this time, but we can examine the patient's disturbance from different perspectives and use the perspective or perspectives that seem most appropriate in selecting interventions. The clinical utility of a particular perspective is understood as more important than the search for the "truth."

If, with this brief introduction to complex causality in mind, we view remedicalization as a current and strong trend in psychiatry, questions can be raised. Although remedicalization or strengthening the identification of psychiatry with medicine in general may make excellent political sense in light of the attack from what I call

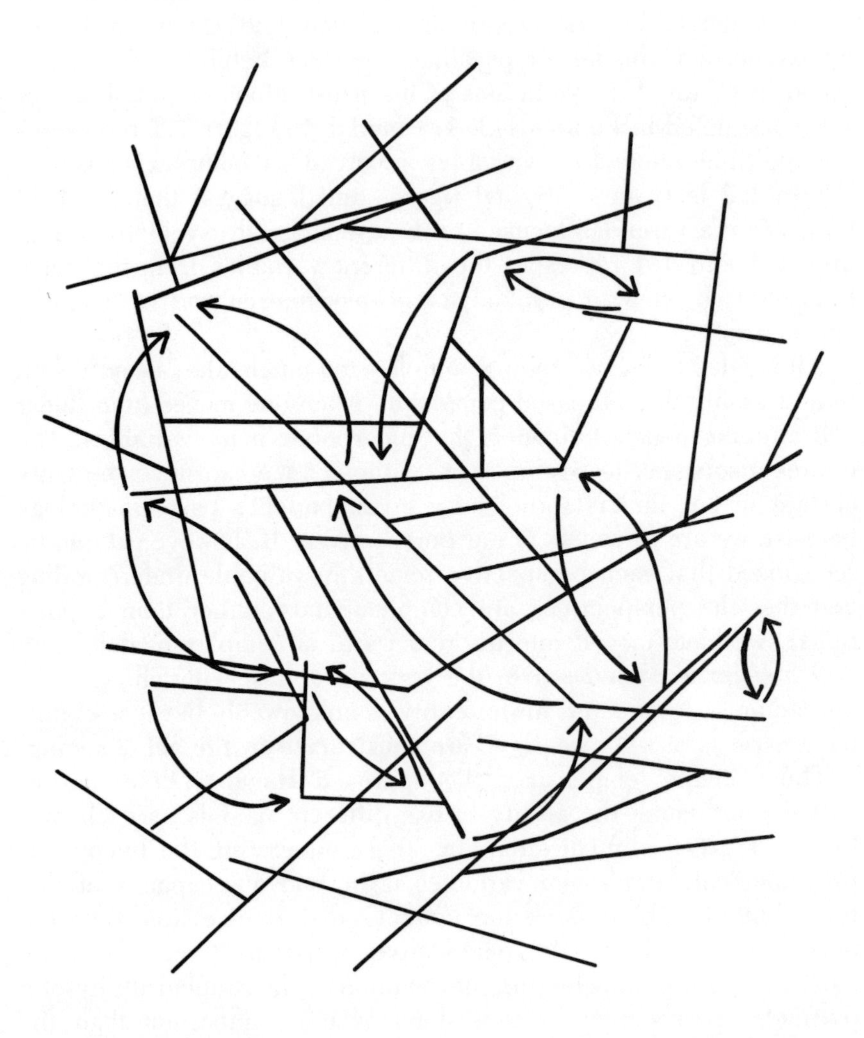

Figure 1.1. Complex Causality. All lines and arrows denote causal relationships. (Modified with permission from: Beahrs, J. O., 1986. *Limits of Scientific Psychiatry*. New York: Brunner/Mazel.)

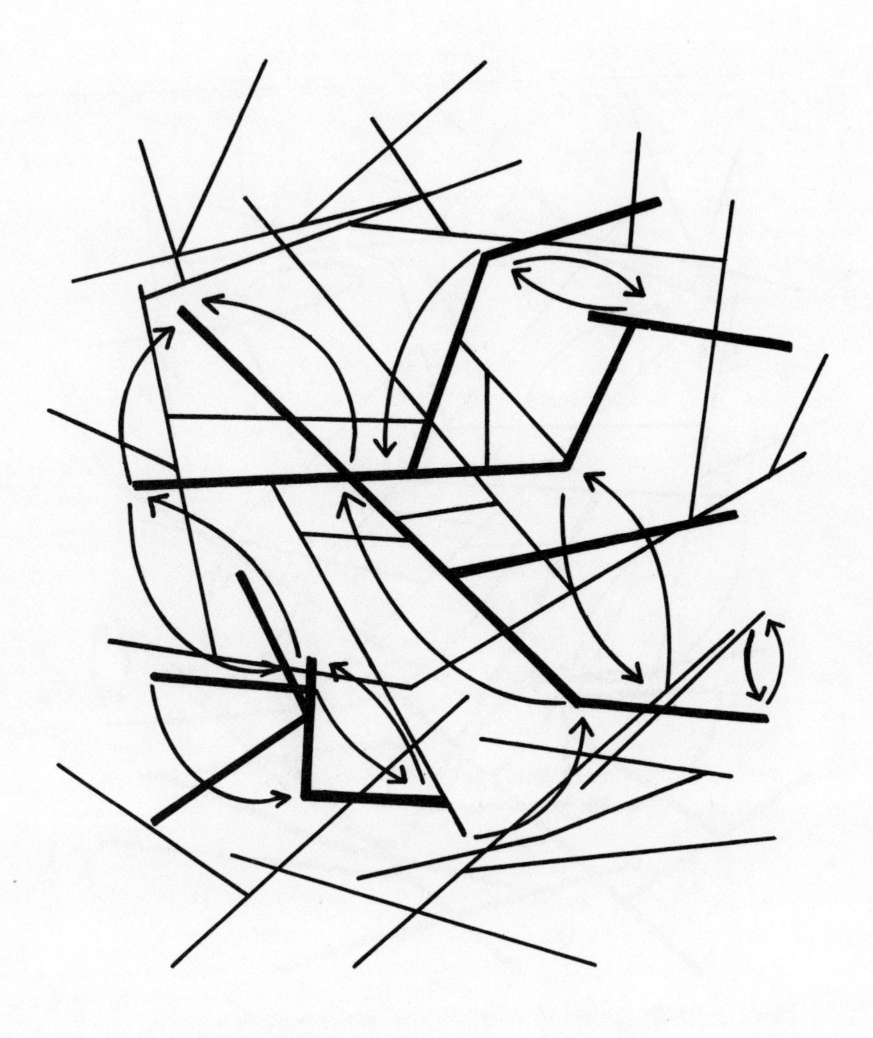

Figure 1.2. Complex Causality. All lines and arrows denote causal relationships. The bold lines are those causal processes focused on by a biological psychiatrist. (Reprinted with permission from: Beahrs, J. O., 1986. *Limits of Scientific Psychiatry.* New York: Brunner/Mazel.)

Figure 1.3. Complex Causality. All lines and arrows denote causal relationships. The bold lines are those causal processes focused on by a psychodynamic psychiatrist. (Modified with permission from: Beahrs, J. O., 1986. *Limits of Scientific Psychiatry.* New York: Brunner/Mazel.)

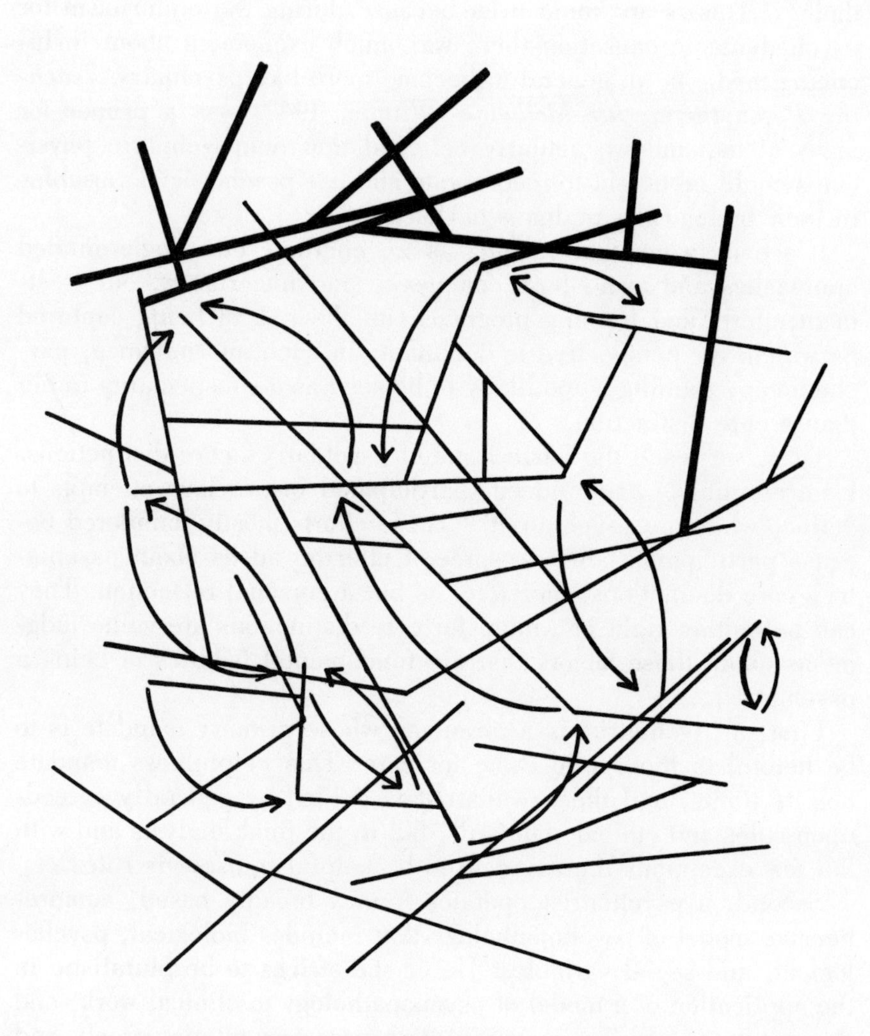

Figure 1.4. Complex Causality. All lines and arrows denote causal relationships. The bold lines are those causal processes focused on by a family psychiatrist. (Modified with permission from: Beahrs, J. O., 1986. *Limits of Scientific Psychiatry*. New York: Brunner/Mazel.)

"radical psychology," it moves us back to a biomedical model of causality and away from complex conceptualizations of psychopathology. This is an ironic twist because during the enthusiasm for psychodynamic causation there was much excitement about influencing medicine in general to become more like psychiatry. *Teaching Psychotherapeutic Medicine* (Witmer, 1947) was a primer for many of us, and we actually believed that nonpsychiatric physicians could be taught to incorporate and use psychosocial variables in their biological paradigms of disease.

It seems to me that as long as we continue our single-minded approaches and avoid both complexity and uncertainty, our treatment efforts and training programs run the risk of being captured by whichever perspective is dominant. In such an ambiance, psychotherapy training is too likely to be seen as a subspeciality rather than a core distinction.

Thus, we reach the basic issue of psychiatry's core distinctions. I am mindful of and, indeed, participated in previous attempts to define "what is a psychiatrist?" These efforts mostly foundered because participants failed to agree. I offer my ideas about psychiatry's core distinctions, therefore, as but a personal reflection. They can be neither right or wrong, for core distinctions are value judgments about those factors that are fundamental features of being a psychiatrist.

First, a psychiatrist is a physician whose primary mandate is to be helpful to those who come for care. This helpfulness mandate has its limits, and the psychiatrist is guided by generally agreed-upon rules and ethical standards, but in the final analysis and with but few exceptions the mandate to be helpful transcends rules.

Second, a psychiatrist operates from a broadly based, comprehensive model of psychopathology that includes biological, psychological, and social variables. He or she strives to be pluralistic in the application of a model of psychopathology to clinical work, and if competent only in one perspective uses consultation freely and refers to others when his or her perspective is not adequate.

Third, a psychiatrist is able to sit with another person, couple, or family to explore the clinical dilemma in a way that facilitates learning. As part of this core distinction, a psychiatrist is knowledgeable about transference and countertransference generally, and his or her own transferential diathesis specifically.

For me these are the keystones; all other skills rest on these three core distinctions. Because these three are my values and because I believe that powerful economic forces, the remarkable advances in the neurosciences, and the difficulties involved in dealing with complex causality taken together threaten them, I wish to share my experiences as a teacher. The format for my teaching is the basic skills of psychotherapy. I hope, however, that what I teach is more than technique and involves something of the complex processes by which a psychiatrist comes to know and treat those who come for help.

CHAPTER 2

Teaching as a Search for Competence

The director of the outpatient clinic gave me the name of a patient in the waiting room, told me to get my patient, go to the newly assigned office, start "doing" psychotherapy, and after the interview he would meet with me and discuss the interview. I had no prior instruction, was anxious, and did what came naturally—took a history. As I asked the patient question after question, I came to relax in the increasingly familiar ambiance of an initial medical interview.

This initial psychotherapy experience remains the norm in many training programs. It is a dreadful way to begin. Imagine training airline pilots in such a way. "There is an airplane. Go fly it, and after you've landed we'll discuss how it went." To rely on such an unstructured learning process is unheard of in other professions. It is much like teaching swimming by throwing a 4-year-old into deep water and assuming that out of his or her frantic thrashings a swimmer will emerge. In addition to its cruelty, it is not good teaching technique, nor are most initial training experiences in psychotherapy. They are not humane and, in light of what is known about the process of learning, they are inadequate teaching.

The continued reliance on such training procedures must reflect many factors. I think of it as a painful rite of passage, an initiation ceremony or hazing, that needs to be abolished. Psychotherapy is a delicate art. Its seeds should be planted carefully and nourished painstakingly. This perspective is even more important today than it was 30 years ago. Then there was no question but that psychotherapeutic competence was a core distinction of psychiatry. Now we debate whether psychotherapy training should even be a part of residency training. If psychotherapy training is to receive less time and emphasis, we should seriously reconsider the most effective and humane way to train residents.

Respected colleagues who disagree with this perspective suggest that good supervision and personal therapy or analysis prepare residents adequately. I do not minimize the importance of either supervision or personal therapy or analysis. They form two legs of the stool, but the third leg is systematic instruction in the processes of psychotherapy.

When I finished training and entered private practice I thought of myself as conventionally well-trained and quickly became very busy. I enjoyed the excitement of my practice, but after only a year or two I began to struggle with some doubt centered about several psychotherapeutic treatment failures. Although most of my patients seemed to make progress, I had difficulty understanding those who didn't. I sought consultations from several senior therapists I admired. They listened and focused on aspects of the patient's psychopathology that I either hadn't understood or had minimized. Each consultation centered on the patient, with very little, if any, attention given to me or to the process of my way of doing psychotherapy.

I turned to a literature review, but found nothing helpful. The few articles I found suggested that patient variables were responsible for treatment failure, and that psychotherapy resulted in either improvement or "no change." No consideration was given to a patient's worsening as a result of treatment!

By tape recording interviews from my practice with patients whose psychotherapy seemed to be going well and with those who were making no progress, I hoped to distinguish differences in the processes. Out of the clear differences between the two types of sessions, one aspect stood out more than others: the affective component of the interviews. Although most of my work at that time was content-focused, I was particularly insensitive to affective cues in those sessions that were from stalemated therapy.

Over time, I came to understand that my failure to move into the patient's affective experience had many roots, but at the time I chose to consider this as involving, most of all, a specific learning deficit of mine that had more negative impact on psychotherapy with some patients than with others. During my training, affective issues had not been discussed in any comprehensive way, and in the absence of recordings that might have made the deficit apparent, I had simply failed to learn important lessons about affect in psychotherapy.

At that time, the residency program at Timberlawn Psychiatric Hospital was reconstituted, and I organized an "affect sensitivity" course for residents.

The course has grown, changed, and is now required, but during the first four to five years its sole concern was affect (known as "Lewis's Empathy Course"). The residents and I read everything about affect and empathy available at that time, but from the start the reading was secondary to experiential exercises. This emphasis reflected my belief that learning by doing is a necessary experience for therapists-to-be and that learning about therapy is helpful, but hardly sufficient.

The exercises we used initially came from the work of Rogers and his students (Rogers, 1961; Traux & Carkhuff, 1967). These consisted of a series of "patient stimuli statements" recorded on tape, to which the residents responded by writing empathic responses. Each statement reflected a specific affect (sadness, anger, etc.) and a specific content (relationship to a parent, relationship to therapist, etc.).

Several thing stood out then—as they do today—about exposing residents to such experiences. First, each resident heard or processed various affects differently. This was particularly true when the patient stimuli statement expressed affects of only moderate or subtle intensity. Second, in their responses many residents avoided labeling affects specifically, using instead general descriptions such as "It is upsetting" or "It is difficult." Third, when the residents were asked to speak their responses to the recorded stimuli statements, many spoke their responses in a flat and detached way. Although focusing on affects, the delivery of the responses was incongruent with the response itself. Finally, even at this early stage of the seminar, many residents did not hear or process certain affects as most others did. Thus, residents were encouraged early on to examine idiosyncratic difficulties with certain aspects of patients' experiences.

Somewhat later in the seminar, a different approach to identifying idiosyncratic difficulties evolved in the form of a Forced Fantasy Exercise. A series of slides involving pictures representing the stages of life from birth to death and involving affective messages such as joy, sadness, anger, and erotic arousal were presented briefly to the group, with the instruction to write one's fantasy about the projected picture. The fantasies are read aloud. Often the fantasies are similar, but several residents produce "deviant" fantasies involving a very different perception and interpretation of the pictures. Thus, a picture seen by several residents as prepubertal children playing on the beach was constructed in the fantasies of one as children running from a dangerous animal. The transformation of play to danger could then be discussed in the group setting (if the resident wished).

These experiences formed another early opportunity for students to begin to understand their individual difficulties in processing certain themes and affects. It is important to emphasize that such idiosyncratic responses are not to be pathologized; rather, interpretations involve the construct that everyone has thematic and affective areas that are processed differently. The task is to identify one's own areas as early as possible in the training experience.

As might be imagined, the amount of self-disclosure and the emphasis on nonjudgmental responses to each other's productions often lead to an intense group process that is supportive in nature. Years after the seminar, former residents have commented on the supportive aspects of the seminar experience.

The use of videotaped patient stimuli statements increases the complexity of the learning process. Residents are exposed to other types of affective signals: facial expressions, posture, body movements, autonomic phenomena, and paralinguistic messages. We read relevant literature and discuss intriguing issues such as incongruence between a patient's words and his or her other affective signals. This leads naturally to the use of confrontation—when and how to intervene in such a way.

Another task soon surfaces: the language of psychotherapy. Here a paramount concern is the therapist's choice of words, use of various grammatical forms such as interrogatives ("How did that feel?"), directives ("Tell me what that was like."), and reflections

("It must have been difficult."), and the use of rhetorical devices. The therapist's deliberateness and spontaneity are explored.

Throughout the seminar experience, I try to avoid an approach emphasizing a "right way"; rather, the emphasis in on the complexity of having multiple options, the factors that may influence selecting a particular option, and the inevitable development of a personal psychotherapeutic style. This approach is difficult for beginning residents who quite naturally wish for a relatively simple set of explicit guidelines. The residents' wish for explicit guidelines leads to what may be considered the last major piece of the first, or affective, stage of the seminar: the introduction of residents to the interface between personal values and psychotherapy.

Relevant literature includes the works of Spiegel and Kluckhohn (1971), Halleck (1974), and, for an appreciation of ethnic considerations, Fabrega and Manning's (1973) writings were most helpful. The major thrust, however, is to encourage the residents to define their own values. The Rokeach Value Survey (Rokeach, 1973) became available, and taking it became a part of the early weeks of the seminar. This survey involves the forced ranking of two lists of 18 values, one termed "terminal" (family security, wisdom, inner harmony, etc.) and the second termed "instrumental" (honest, loving, obedient, etc.). Each resident's value rankings are compared with those of colleagues and with published national norms. Two conclusions became inescapable: although individually the residents are much alike in value orientations, there is a minority of residents who are much less like their colleagues, and, taken as a whole, the residents are very different in their value orientations than the national norms.

This can be illustrated by the residents' responses to the terminal value of salvation. Although the majority of residents are Protestant, with Jewish and Catholic representing one-fourth of the total, on the average and through the years, *salvation* is the last (18th) ranked terminal value. Despite this low ranking, for a small number of residents *salvation* is ranked first. In the national norms, however, salvation is a midranked value (8th or 9th).

This and many other differences between residents' value rankings and national norms introduce the residents to the complexities of treating patients who may hold dear values very different from their own. How to recognize this difference, how to avoid pathologizing it, and what options one has if the differences in values interfere with the capacity for a treatment alliance are, along with other complexities, discussed by the group.

I have no proof of the impact of these early years of the seminar—the affective phase—on my work as therapist. I believe, however, that it had a major influence. Perhaps the changes I experienced in myself would have come about in any event, but no longitudinal, empirical data clarify what, if any, predictable changes occur in the maturation of psychotherapists. Three changes in both my work and my experience of myself are clear to me.

The first change was that my work became much more focused on patients' affects. Although in some situations focus on affect is to be minimized or even avoided, overall my tendency has come to search for the "deepest" affective level that the patient and I can achieve. The focus on affect has had many reverberations on my interests. To mention but a few of these interests: empathy, resistances to affective exploration, right cortical influences on affective reception and expression (the aprosodias), alexithymia, the role of affect-focused responses on narrative flow, and affects and the regulation of interpersonal distance. These become major areas of interest in my psychotherapeutic work, teaching, and supervision.

A second change in my psychotherapeutic work involved a movement away from directive approaches to a more collaborative orientation. I dislike acknowledging the extent of my early directiveness. If asked, I would have said (and believed) that since psychotherapy of a psychodynamic type was collaborative, and since that's what I did, my work was clearly collaborative. It was only when I undertook the task of teaching residents a more collaborative approach and, thus, had to deal with the relationship of what a therapist actually says during psychotherapy to the dimensions of directiveness and collaboration that it became clear that in my reliance on the interrogatives I established and controlled (hence directed) the interview.

Although some clinical situations call for directive approaches, much of the everyday work of psychotherapy is of a collaborative nature. Only by paying attention to the actual dialogues of one's therapy, however, can the therapist decide if he or she is collaborative or directive.

A third change in my way of working with patients, given impetus by the early years of the seminar experience, involved a much greater attention to the here-and-now of the psychotherapeutic session. Although I had been taught to be attentive to signals from patients suggesting feelings about me, my mind was filled with injunctions: "Don't interpret transference until it presents as resistance." "Interpret negative transference before positive transference." Thus, though I was attentive, my focus was on the patient's feelings, and, for the most part, I said little.

In teaching the seminar, and especially in reviewing videotapes of residents' role-playing exercises or later interviews with actor-patients, the focus on what the patient reported feeling was augmented by a concern with therapist-patient interactions. This is the interpersonal ballet in which what one participant does is both a response to the other's immediately preceding action and a stimulus to an often easily recognizable response from the other. This focus—so clear on the videotapes—received additional encouragement from the interactional focus of marital and family research, another major part of my professional life.

It was not just a question of changing the focus (perhaps "adding the interactional focus to the previous patient-centered orientation" is better description). Rather, it also involved coming to realize that clarifying in a collaborative way with the patient, an interactional pattern that is occurring right here and now between you and me is often the preferable intervention. This involved a major change in the management of what I earlier thought of as transference. Because I am not a psychoanalyst and see most of my individual psychotherapy patients once or twice a week, I do not want a full-blown transference neurosis to evolve. Indeed, in my early years of practice, when such transference did occur (often with patients with fairly "primitive" psychopathology), the results were rarely helpful and often messy.

The way to avoid such phenomena—at least as much as possible—is to keep what is going on as explicit as possible. Thus clarification of here-and-now interactional patterns has the advantage of helping patients not only to understand better their own interactional tendencies, but also to learn something of their impact on others. Unless there are clear reasons not to use such an open approach with certain patients, I do so with regularity.

Although teaching the seminar changed my understanding and approach to my own psychotherapeutic work in other ways, the increased focus on affective messages, movement toward a greater level of collaborativeness, and increased emphasis on the here-and-now of the psychotherapeutic relationship are changes that spring most clearly from the early teaching experience.

The second phase of the seminar evolved naturally from the early emphasis on affect to the more cognitive interventions. The emphasis on affect and empathy needed to be balanced by an equal focus on the roles that occupy much of each psychotherapy session: the psychotherapist as observer, data collector, synthesizer, model-builder, and deliberate intervenor. These activities are work done at some metaphorical distance from the patient, not in order to enter his or her experience empathically but to understand the patient's psychopathology and to plan interventions.

For most residents, this detachment was less anxiety-filled than the earlier empathic exercises, having less to do with encounter and more to do with sleuthing and with diagnosis, with something more "objective." For me, the challenge was to devise effective ways to teach these fundamental cognitive psychotherapeutic processes. Gradually, a list of processes evolved, and now, some 15 years later, I describe the list in the following way.

1. *Processes that facilitate narrative flow.* Here, in addition to the role of empathic reflections in facilitating the patient's narrative, I emphasize what Havens (1986) calls projective statements ("*It's as if they might have known. . . .* "), the use of general encouragers ("*Go on*" or "*Help me understand that better*"), and the use of brief

summaries ("*You say it was after the fire that you first noticed her fears and then you, too, began to withdraw*"). Although there is some selection by the therapist, the patient's narrative is allowed to flow as freely as possible. The therapist minimizes direct questions and molds the narrative as little as possible.

2. *The recognition of repetitive patterns in the patient's narrative.* Here a variety of patterns are attended such as, for example, the predominant affects that punctuate the narrative, the major themes of the narrative, including repetitive patterns of interpersonal relationships, the emergence of a central dynamic conflict, and the pattern of mechanisms of defense that recur throughout the narrative.

3. *The ability to make an initial formulation of the patients dilemma.* This inductive task needs to be introduced to residents in a relatively simple manner. To be able to identify major affects, the central dynamic conflict, repetitive interpersonal transactions, and perhaps several defense mechanisms is a considerable task at the beginning of the training experience. As the seminar progresses, increasingly complex formulations can be introduced, but at the beginning it is best to emphasize only those features that are clear and require minimal inference.

4. *The use of common cognitive interventions.* Here the emphasis is on confrontation, clarification, and interpretation. Most beginning residents have the notion that such interventions are delivered with solemn authority, a brilliant bolt of lightning from above, something you do *to* rather than *with* patients. I came to understand that the task is how to use these often powerful interventions within a collaborative relationship or, in other words, how to involve the patient in as much of the work of the therapy as is possible. These interventions are rarely single statements; rather, they are interchanges in which the therapist, often with considerable tentativeness, may begin the joint effort.

"*It seems as if there is something familiar here.*"

"*Oh?*"

"*Do you have some sense of it?*"

"You mean, well . . . it reminds me of, somewhat of, what we discussed some time ago . . . about, well . . . my father going away. . . ."

"Always being left by someone."

"Yeah . . . yeah. Actually, either someone leaving me or my leaving. Like when I went off for that special course. It's the leaving that seems to get me going."

In this example, a reasonable chance exists that the patient will feel that the expanded understanding results, at least in part, from his or her own efforts, that it is not a gift from a powerful other. As such, the insight is likely to have more meaning.

The learning exercises in this stage of the seminar rely mainly on observing videotaped interviews. Transcriptions of interviews are also helpful. The general format is to assign each resident a particular observational perspective. One resident may be asked to identify major affects, another to report on the therapist's use of interventions, and others to note various aspects of the patient's presentation, the therapist's behaviors, and the evolving patient-therapist relationship. Each resident reports his or her observations to the group, and the findings are discussed. Later in the seminar, all residents are asked to take all the observational perspectives, and one is chosen to make a full report of the interview.

Initially we use videotapes of experienced therapists' interviews, my own and others. It is important to be open about mistakes, missed opportunities, and other less-than-optimal therapists' responses. As Kramer (1989) notes, therapy is the creation of a context in which mistakes are useful.

During this stage of the seminar, actors are employed and given brief situations and told to improvise. ("You are a drama student and your girlfriend has left you. You are sad and thinking about suicide. Go wherever the interview takes you" or "You are a staunchly prolife college student who finds herself pregnant and unable to tell your very religious parents. You don't know what to do and have come to the student health center. Go where the interview takes you.") Each resident does a 20-minute videotaped interview with each actor. These tapes are reviewed by the class, with each viewer having an assigned observational perspective. The residents are quick to note that some interviews seems to go better than others; that many interviews involve complex value-loaded issues; and that experiencing considerable anxiety pulls one back to doing a more directive inquiry.

Although describing the seminar in stages is helpful, there are no clear demarcations. Rather, the emphasis changes as different perspectives are added. At this point in the seminar, the residents have been exposed to and experimented with two different sets of processes: those having to do with affect, empathy, and the attempt to move close to or into the patient's experience, and those having to do with the more detached, cognitive observations and interventions. They are thus introduced to the therapist's movement toward sharing the patient's experience, then a few moments later the movement back to a more cognitive, analytic perspective of their ongoing interaction.

They are introduced to this figurative movement as a reflection that life itself is experienced in these two very different ways. One state is called connected, fused, attached, merged, I-thou, and the other is called cognitive, separate, differentiated, distant, I-it. The residents are then introduced to the idea of monitoring their own movements from one state to the other.

These years were so filled with home and hearth, adolescent children, the deaths of my parents and sister, helping to build clinical programs at our hospital, initiating research at our foundation, private patients, making new friends at the national psychiatric level, exchanging experiences and ideas with those friends and close colleagues at Timberlawn, and many other events and processes that it is impossible in any definitive way to isolate the impact of one event or process. Yet, somehow, my seminar with the residents and, by now, other seminars for practicing psychiatrists, clinical psychologists, social workers, nurses, clergy, executives, and others seemed to have a major impact. Always there was the need to move from a major focus at the level of the abstract—learning about narrative flow, empathy, and confrontation—to a move toward the more concrete—the details of the interview, the complexity of moment-to-moment decision-making when faced with multiple options.

I was also supervising others' therapy of patients four to six hours each week. I asked each resident or clinician for a brief summary *plus* an audiotape of each session. Sometimes we listened to a segment requested by the supervisee, more often to a randomly selected segment. Always we were preoccupied with the data of the interview. *"What is the patient saying?" "What do you think might have been involved in your switching the topic?" "What are the possible meanings of this sequence of mutual interruptions?" "How do these data fit with your initial formulation?" "Were you aware of any fantasies you had during or after this session?"*

These and other issues captured our attention. This preoccupation with details and process came, of course, to influence my own psychotherapeutic work. All therapists are to one extent or another their own supervisors—there is nothing unusual about this. It was my increasing concern about the details* of the session, in some ways a more exacting focus, that appeared to grow out of the seminar teaching and the increased supervision.

A second effect of my teaching involved a deeper exploration of myself. The observation of the figurative movements toward and away from the patient's experience brought me into closer contact with my propensity to, and comfort with, the detached position. Almost without exception residents share this propensity, and such a proclivity is overdetermined but certainly strengthened by the experience of medical school.

A third reflection of the impact of the seminar on my work as a therapist involved the deepening sense of adventure in each interview. As I asked residents to learn a new form of exploratory interviewing, to move away from depending solely on the highly structured directive interview, to view each session as an adventure, the exact ending of which is unpredictable at the outset, I became more interested in the adventurous component of my own psychotherapeutic work. This involved a different quality of curiosity, excitement, and, gradually, a greater interest in the idiosyncratic features of each patient's experience. It became more apparent that my earlier work had involved a greater emphasis on the search for commonalities, those features of a particular patient that he or she

*After writing this, I read Safire's (1989) interesting column exploring unsuccessfully for the origins of the phrase, *"The devil is in the details."*

shared with other patients at some level of description. Although those shared characteristics are central to descriptive diagnosis and, hence, not without clinical relevance, they are of but limited value for the process of psychotherapy.

I have the sense that as these self-observations are read, the reader will exclaim, "*Why, of course, didn't you always know that?*" The answer is "*No, I haven't always known that.*" Even if these issues were present in the work of those authors whom I have experienced as most instructive about psychotherapy, they took on a different meaning for me and my work as I faced the challenge of translating their written work to exercises for beginning therapists.

The third stage of the seminar involved a growing need to teach something of relationship structure. Relationship structure, a central construct of the systematic study of marital and family systems, is defined as the more-or-less enduring patterns of interactions in an interpersonal relationship. Structure in this sense is an interactional process with a very low rate of change. Relationship structure is thus concerned with how a couple relates—the repetitive patterns of their communication—and is less concerned with the content of communication. It is impossible for two people, for example, to have an ongoing relationship without their having a relationship structure evolve, negotiated subtly by the participants, negotiated most often without awareness.

When a relationship structure is in place, however, it exerts a powerful influence on what goes on between people. The key issues around which the structure is crystallized are power and distance. The issue of who is to be in charge of distance-regulation, as well as who is to be in charge of other relationship issues, is paramount. They are introduced into the seminar because of their relevance for understanding the psychotherapeutic relationship. In essence, I have taken concepts central to the study of marital and family systems and applied them to the study of the therapist-patient relationship.

Interpersonal power can be studied at different levels, but for purposes of the seminar the focus is on those more apparent and clear levels requiring less inference than does "deeper" analysis. Further, if one starts with the assumption that the culture gives the physician or other therapist considerable sanctioned power, the therapist's goal in many forms of psychotherapy is to behave in ways that minimize the power differential between therapist and patient, to strive to make the relationship as collaborative as is consistent with the goals of the therapeutic endeavor.

Sometimes, however, the patient may present a formidable challenge because he or she appears determined and skillful in establishing an extreme relationship structure in which he or she is in complete control, or else is helpless and submissive (passive control), or in which conflict about this central relationship problem is ever present. For some patients, much of the work of therapy is completed when a more reciprocal and collaborative relationship structure is accomplished.

Bateson's (1972) concept of "command" messages, those that intend to influence the relationship structure, is helpful. The message is, "*This is how I want you to treat me, and this is how I want to treat you.*" Residents are encouraged to listen for the affect, content, and command messages in patients' statements.

In addition, interpersonal systems research identifies mechanisms that point to who has the power in a relationship: attempts to control the topic of communication, frequent interruptions, topic changes, and repetitive questions.

A careful analysis of patient-therapist interchanges in videotaped interviews is useful material for the study of relationship structure. It is helpful to start with examples of experts. A tape of a rational emotive therapist interviewing a patient is contrasted with an interview by a Rogerian therapist. These interviews reflect very different approaches to the issue of power and distance. The rational emotive therapist often dominates the patient, interrupting frequently with statements like "*What you really mean is. . . .* " The Rogerian often seems more passive, nondirective, and primarily interested in helping the patient to clarify his or her experiences.

Although the use of the tapes of experts is a helpful beginning, the videotapes of the residents' interviews are the major source of teaching material. Those with actor-patients are followed by interviews with actual patients. Often, one resident asks one of his or her hospitalized patients to consent to be interviewed by a colleague who does not know the patient. The videotaped interview is then analyzed by the group from the multiple perspectives noted earlier, but, as the concept of relationship structure is elaborated, with considerable emphasis on this perspective. Residents begin to recognize their own and their patients' attempts to structure a particular kind of relationship.

Particularly instructive and exciting are times when the participants are able to identify bits and pieces of a patient's description of a current conflictual relationship, a similar relationship structure from childhood, and the patient's subtle efforts to construct the same type of relationship with the interviewer. The obvious implications of such observations and hypotheses for the development of transference distortions come to life on the screen.

The residents are also introduced to Kagan's (1967) Interpersonal Process Recall Technique and its relevance for understanding relationship structure. In this technique, each participant in a videotaped interview is interviewed (also videotaped) by a third person while watching the initial interview. By focusing on sudden changes in the flow of the initial interview, the third person attempts to get at the inner experience of the original participants at points of change. Kagan's method helps with understanding the participants' movements toward and away from each other and the anxiety inevitably experienced by both interviewer and patient whenever each participant's threshold for closeness or distance is exceeded.

At the point in the seminar when the residents have become familiar with the concept of relationship structure, we are well along toward the final stage of the yearlong seminar. By this time, the residents have begun to see patients in psychotherapy and are being seen by supervisors. Often (and with their supervisor's permission) they will bring tapes from their own psychotherapeutic efforts. A level of openness about their work is in the process of becoming established; anxiety about mistakes diminishes; personal areas that need attention are often clearer.

> The impact on my own work of teaching this aspect of the seminar is tightly interwoven with the marital and family research that has occupied much of my time for over 20 years and out of which my interest in relationship structure began. Starting with an interest in treatment evaluation and a research project exploring the outcome of hospital treatment for

severely disturbed adolescents came the need to explore the impact of family variables on treatment outcome (Gossett, Lewis, & Barnhart, 1983; Lewis, 1989; Lewis, Beavers, Gossett, Phillips, 1976; Lewis & Looney, 1983). Out of the need for a control group of families containing well-functioning adolescents came the interest in healthy family systems. Out of the study of healthy families came a focus on the parental marriage. Out of an appreciation of the importance of relationship structure in marital relationships came the attempt to test its usefulness in other relationships.

The preoccupation with relationship structure and, in particular, the early stage of relationship formation during which the underlying negotiation is most prominent; the hundreds and hundreds of hours spent watching the videotaped interactions of couples attempting to solve experimental problems; the use of rating scales that focus attention on particular dimensions of the relationship structure—these and other factors inevitably led to a greater self-consciousness about the types of relationships I establish, most notably with my wife, but also with my children, close friends, and patients.

How much this preoccupation actually changed me and changed my propensity to try to establish a certain type of enduring relationship structure is open to question. From my perspective, the change has been clear, and some confirmation is found in the observations of those close others who have known me for years.

The direction of change has been alluded to in earlier sections. I have tried to move in the direction of greater collaboration, an increased responsiveness to affect—that of others as well as my own—and an increased capacity for closeness in both my personal relationships and my professional work.

Stage four of the somewhat artificial division of the seminar into stages involves an emerging focus on the subjective responses of the therapist and, most importantly, on how to use them in the service of effective psychotherapy. For beginning therapists, this aspect of the seminar is so fraught with anxiety regarding self-exposure that it must be dealt with slowly and cautiously.

The medium for introducing this important component of the therapeutic process is the fantasy. The Forced Fantasy Exercise, described earlier, is introduced during the initial months of the seminar. The evocative nature of the projected pictures is sufficiently strong to facilitate the emergence of a consensual fantasy. Although there is concern about reading out loud a fantasy different from those of one's colleagues, the use of evocative pictures offers some protection—one is not completely on one's own.

The videotaped interviews that are such a large part of the latter months of the seminar are a second source of learning to use one's subjective responses. Here, two procedures seem particularly helpful. One is to note sudden shifts in the interviewer's behavior—change in posture, a particular body movement, a mispronounced word, an "out-of-the-blue" topic change, and other such signals that alert the viewers to the possibility that the interviewer is struggling. The tape is stopped and the resident involved is asked if he and she can get back in touch with that moment, in touch with some thought, feeling, or fantasy that was, perhaps, discomforting. This is often productive and the resident-interviewer is able to articulate what was entering his or her mind.

In addition to the focus on sudden behavioral changes during the interview, a resident other than the interviewer is assigned the task of constructing a fantasy involving himself or herself and the patient in the videotaped interview. These fantasies (and the group acknowledges that some of those that are read aloud are second or third fantasies rather than first fantasies) are responded to by the other participants, with particular emphasis on the insights provided regarding the patient's ability to evoke certain feelings and fantasies in others.

The reference to first, second, and third fantasies speaks to the acceptance of ever-present defenses regarding self-esteem issues. I say to each group that whether or not one chooses to read his or her first fantasy, one knows what it is and can work privately on understanding it and using it.

Throughout this stage of the seminar, the emphasis is on the inevitability of having subjective responses to all patients. Such responses are a very important source of potential understanding of the patient, the treatment process, and one's self. Thus, subjective responses are of special concern only if the therapist experiences a growing sameness in his or her subjective responses to all patients or to all patients of a particular gender or age. This suggests that the subjective response reflects something within the therapist that needs immediate attention.

A second source of concern is the temptation to act on the basis of one's subjective response. The act itself may vary widely (e.g., inappropriately prescribing or not prescribing a psychotropic agent, increasing or decreasing the frequency of appointments without clear indications), but invariably the subject of sexual feelings about patients emerges and becomes a central focus. In agreement with Kramer (1989), I have taught for years that there is something wrong with the therapist who doesn't have sexual feeling or fantasies about some patients. The simplest presumption is that the therapist is holding the patient off at a distance because the intimacy involved in working closer to the patient's experience may facilitate sexual arousal.

Residents are taught the inevitability of sexual responses to patients does not mean that they are meaningless in understanding a particular patient, one's self, or the process of the relationship. Sexual feelings may emerge because of the intimacy involved in the relationship, but those feelings may serve many purposes, including defensive ones.

This aspect of the seminar is a good place to introduce the residents to current knowledge about what one should do if it appears that control of one's subjective response is waning. The urgent need for consultation is emphasized; the clinical data about the harm that results from actual sexual contacts are discussed; and both the legal and ethical ramifications are introduced.

Throughout this stage of the seminar, I use frequent examples from my psychotherapeutic experiences. I try to be open about subjective responses to patients in videotaped interviews, use as teaching vignettes summaries from work with de-identified patients, and often share dreams that I have had about patients in order to illuminate subjective responses. All of this reflects a commitment to openness, sharing, and the quest for reduced defensiveness. I try to model this throughout the seminar, but repetitively suggest that success is, at best, incomplete. The power of the need to avoid, deny, repress is strong, the underlying fears too great to expect complete victory.

I trust that the message in all of this is clear to the reader: First, that the search for psychotherapeutic competence is a lifelong journey; second, that one should use all available vehicles in that search: supervision, personal therapy or psychoanalysis, the published and recorded work of experienced clinicians, and, for the fortunate, the opportunity to teach the next generation with the mandate to operationalize and make learnable the process we use in our work with patients.

Other messages are intended. For educators in psychiatry, the message is that we have not been creative enough in searching for ways to teach processes central to our craft. The seminar briefly described here is but one effort. It could be vastly improved and has been in many training centers, molded by the interests and personalities of those who lead. It has been exciting for me to incorporate into my teaching these creative approaches developed in other settings.

Often residents are left to struggle alone in the beginning of their search for competence. Such is not necessary; there are better ways to get them started.

CHAPTER 3

The Dynamics of Exploration

That there are several different forms of useful clinical interviewing comes as a distinct surprise to all but a few students. Early in a seminar focusing on psychotherapeutic process, a first-year resident takes the role of a depressed medical student and I demonstrate two of these forms in two 4-minute interviews.

The first interview is the familiar directive inquiry. The student-patient defines his or her reason for seeking help as sadness or another "chief complaint" suggesting depression. Through a series of direct questions exploring the well known dimensions of a depressive syndrome, the presence of such a syndrome is established. The flow of the interview comes, for the most part, from the interviewer's knowledge of such syndromes, and it is clear that the interviewer is in charge of the focus and direction of the interview process. In an ambiance of clinical detachment the interviewer subtly invites the student-patient to assume the same stance toward his or her experience. The 4-minute role play is comfortably familiar; the scene might be from any doctor-patient initial contact or, for that matter, from the initial encounter between any professional and client. A typical interview segment follows:

"I'm here because I feel kind of down."

"When did you first notice feeling down?"

"About a year ago."

"What was happening in your life then?"

"Nothing different. My wife and I separated."

"The down feeling started then?"

"Yes."

"And do you have trouble sleeping?"
"Yes, it started several months ago."
"And loss of appetite?"
"Just the last few weeks."

This form of interview is almost entirely concerned with those aspects of the patient's experience that correspond to definitions of syndromes learned by the interviewer. The patient's experiences are organized by the interviewer's set, which emphasizes careful descriptions, temporal relationships, and the search for the similarity of the patient's experience with that of diagnostic criteria. The interviewer is in charge as illustrated in the decision not to respond to the separation issue but to change the subject to the onset of the sleep problem.

After the class discusses this interview, the same resident is interviewed again with a very different interview structure, "collaborative exploration." A segment of this type of interview follows:

"I'm here because I feel kind of down."
"It is a deep, down feeling . . . "
"It really is."
"Nothing feels good."
"My wife left."
"Down and all alone . . . "
"Its always been that way. Alone. I've been alone all my life. Even as a kid I never felt close to anyone."
"For so long its been all alone."
"I'm afraid it always will be. . . . "

In this form of interviewing, the focus is on the exploration of the patient's affect in the here-and-now. The interviewer, avoiding directly shaping questions, responds to the patient's statements with reflections. There is less interest in symptoms and diagnosis. Rather, there is a facilitation of the beginning of the patient's narrative which, in this example, appears to involve depression and the experience of aloneness. Narrative flow comes more distinctly from the patient's experience. The interviewer is not detached, but communicates a sense of understanding and feeling. On some occasions the interviewer comes to feel what the patient is experiencing, and early memories of his or her own may filter into consciousness. Momentarily there is a connection, a period of much reduced interpersonal distance.

The class discusses this different form of interviewing, and the students compare the two different interview structures. A list of characteristics that differentiate the two structures is written on the blackboard. I emphasize how firmly ingrained the directive form of interviewing is in most of us, how familiar and comfortable it is, and how well suited it is for some circumstances but not for others.

In the next several sessions of the seminar, the residents are assigned partners and role-play doctor and patient alternately. They attempt a 4-minute collaborative exploration and their efforts are videotaped. Each videotaped role-play is dissected by the instructor and the other residents in the seminar.

Most residents find it extremely difficult to avoid lapsing into a directive inquiry or taking a traditional history of chief complaints. If the beginning of the role-play is more affect-oriented and reflective, the resident soon finds that his or her uncertainty and anxiety results in the conversion of the dialogue into an inquiry filled with questions through which the interviewer takes charge of the direction and flow of the process.

Following these role-playing experiences, the class members discuss the difficulties they have encountered: the new and strange interview structure, the reduction in anxiety afforded by returning to the more familiar directive approach, the continuing influence of the directive style of interviewing learned in medical school, the comfortable distance from the patient provided by the directive process, and the experienced need to arrive at a formal diagnosis.

During these discussions, questions about the nature and amount of information obtained by each form of interview structure are raised by the group. Most often, these questions reflect concern that "something important" will be missed without direct questioning. They are told that such questions are not out of order in a collaborative interview; rather, their dominance of the interview process is to be avoided. The quality and nature of the data are usually determined by the context of the interview. If, for example, the context demands a diagnosis in a relatively brief contact, a directive inquiry is often the better choice. For many forms of psychotherapy, however, a more reflective collaborative exploration is almost always indicated.

The importance of the context is illustrated by my experience with a middle-of-the-night viral pleurisy with severe dyspnea. My internist did not reflect that I seemed to be frightened, but rather asked a few direct questions about the nature of my pain and whisked me off to the hospital. The context, including the apparent emergency nature of my experience, determined the approach. In most of the clinical encounters in psychiatry, however, it is better to start with a collaborative exploration.

Interviews of a collaborative nature videotaped as part of inpatient consultations also amplify the nature of the information that can be obtained. Assigning each resident a monitoring task is of much help in this regard. Several residents monitor the patient's major affects, other residents monitor the major narrative theme presented by the patient, and other residents monitor the interviewer's use of reflective techniques such as empathic statements, general encouragers ("Go on," "Help me to understand that"), and brief summations. In this way the residents begin to understand the central idea of facilitating the narrative flow and the kind of data that emerge from this type of interview structure. The directive inquiry, learned in the first days of their inpatient assignments, comes to be understood as a very different type of interview structure.

That which becomes apparent to the residents is a very simple fact: The way one talks to another person is an important determinant of the type of relationship structure that will be formed. Directiveness fosters the development of a relationship in which power is clearly held by the interviewer and considerable distance is established between the participants. Collaborativeness allows a less distinct power differential and less interpersonal distance. The directive inquiry looks for symptoms and their temporal sequences. A collaborative exploration seeks to facilitate narrative flow with as little interruption as possible in order to note associative patterns and redundancies.

The two very different ways of interviewing are richly illustrated in Coles's (1989) descriptions of his early experiences with two supervisors. One supervisor emphasized probing questions in order to formulate the patient's dilemma as quickly as possible. The language of that supervision was abstract and psychodynamics was the vocabulary. The other supervisor encouraged Coles to help patients tell their stories, to listen, and to interfere as little as possible.

From the second supervisor's perspective, the concrete details of the patient's life are at the center of one's interest.

Gay (1988) reports that early in his practice Freud was told by a patient, Emmy Von N., that his insistent questions—"*asking her where this or that came from*"—should stop, and that he should "*let her tell what she had to say.*" Listening thus became a method, a road to knowledge mapped out by Freud's patients.

Havens (1973) explicates the perspectives or schools of psychiatry from which different interview structures arise. The directive inquiry is the hallmark of the objective-descriptive perspective. This perspective holds central the traditional idea of illness—as a process or entity of a different sort or class than the patient. In this sense, illness is very real, with distinct boundaries and a typical developmental history. The central goal of the directive inquiry is to find the illness, to make a correct diagnosis in order to establish the category in which the illness belongs. Ruthlessness can be condoned because the interview structure is directed at finding the illness. The interviewer hopes to elicit brief, clear answers: preferably "*Yes*" or "*No.*" Unfortunately for the patient, all that is unique and highly subjective is often seen as of little value.

In the directive inquiry, much attention is given to the patient's chief complaint. It is from the chief complaint that the interview flows, and the interviewer's questions clearly reflect his or her early hypothesis about the nature of the illness.

> Watching a really skillful physician interview a patient using a directive inquiry is like seeing a work of art. I was very fortunate as a medical student to drop out of school for a year to be Tinsley Harrison's research fellow. Harrison, one of the last group of great bedside clinicians occupying academic chairs, selected one medical student each year to be his clinical fellow. This involved having a small office next to his and working with him on a clinical research project. It allowed the medical student to have several hours of direct contact with him most days. During my year with him, he completed the first edition of *Principles of Internal Medicine* and one of my assignments was to read the galley proofs, making me the first medical student to read the text that quickly became the all-time best seller in the history of medicine (Harrison & Beeson, 1950)

Harrison was a master interviewer. Although kind and gentle with patients, his interviews stirred visions of a careful assault on a fortress or of a remarkable chess game. After taking the history and doing a bedside physical examination, he dictated his clinical note, and only then looked at the laboratory findings and x-rays. His emphasis on history-taking, he once told me, was related to our young medical school's mission to produce primary care physicians for rural Texas where, more often than not, there would be little in the way of sophisticated diagnostic equipment. "*The most important medical instrument,*" he said, "*is the pencil.*"

As must be clear, Harrison was the major mentor of my career as a physician. My identification with him was intense, and although he was openly disappointed when I opted for psychiatry, we remained close until his death. Even now, when I watch an experienced clinician undertake a directive inquiry, I think of "The Chief," and when a patient complains to me of headache, weakness, dizziness, or many other symptoms, what I call a "Harrison cassette" snaps into place in my cortex. I often refrain from using these cassettes at the time because they would interrupt the patient's narrative, but they are part of me and still find clinical utility.

For all of the warm memories and admiration of a skillful directive inquiry, a well-done collaborative exploration is, for me, a much more exciting experience. There is a clear sense of adventure, a "what-is-going-to-be-revealed" atmosphere. The goal is not primarily diagnosis, but the facilitation of narrative flow with as little molding and shaping by the interviewer's preconceived set as possible. Further, a collaborative exploration attempts to facilitate the narrative at the deepest level of affective meaning that appears possible for the patient. Although for a few patients such an intense interview experience may not be best, for the majority it is the form of interview best suited to a number of purposes. It fosters a collaborative alliance, allows recurrent patterns of affect, dynamic conflicts, and central relationship patterns to become apparent, sheds light on the patient's relationship to his or her own experience of self, and tests suitability for most psychodynamic psychotherapies.

Beginning residents need to develop a handful of basically reflective techniques which enable them to consider three aspects of each of the patient's statements: affect, content, and what Bateson (1972) calls the "command" message. The last are those aspects of a person's communication that appear to be attempts to structure the relationship with the interviewer along certain dimensions.

For a number of reasons the seminar starts with a major focus on hearing and processing the affective component of the patient's communication. First, most beginning residents attend almost exclusively the content of the patient's communications, and they need to develop a better ear for affect. Second, responding to patients' affective messages may facilitate and deepen the narrative flow. Third, focus on affect often leads patients to feel more thoroughly understood and, hence, willing to risk greater self-disclosure to the interviewer. Fourth, one of the major schools of psychiatry, existential psychiatry, emphasizes the curative impact of "being" deeply "with" the patient in his or her feelings. Fifth, a focus on the patient's affective messages often clarifies the nature and intensity of the patient's resistances.

The focus on affective messages leads us into the arena of empathy. I believe there is a continuum of empathic responses ranging from the acknowledgment of another's feelings (cognitive empathy) to actually sharing another person's feelings (affective empathy). Beginning students are focused on the less complicated and less threatening cognitive empathy. In this form of empathic response, the interviewer remains at some distance from the patient and his or her experience. I emphasize empathic statements that are third-person impersonal, such as "*It is terrible*" or "*It is so sad*" because, as Havens (1986) suggests, this grammatical form suggests the possibility, at least, that the interviewer has or can experience the affect. "*You are sad*" or "*You feel terrible*" does not contain that suggestion.

General encouragers are a second reflective technique. Here the emphasis is on encouraging narrative flow quite directly. "*Please go on,*" "*How interesting,*" and a variety of other verbal and nonverbal expressions indicate the interest of the interviewer and his or her wish to hear more.

Brief summaries are another reflective technique. Perhaps after a moment of silence, the interviewer gives the patient a summary of the salient features of what the patient has just completed, saying, "*So it was after your daughter was born that you first felt anxious and began to notice the recurrent dream of your sister's death. . . .* " Here, as in other simple reflective techniques, the effort is to follow closely the patient's narrative line, to introduce nothing truly new and, most of all, not to change the subject by taking the interview into a different direction. Although it is patently impossible not to select certain affects and themes that clearly reflect one's ideas of what is most relevant, one's goal is to facilitate the patient's story with minimal molding by one's preconceived notions. One hopes to find what is unique about the patient's experience rather than, as with the directive inquiry, to search for symptoms that lead to a syndromic diagnosis—that is, those features of a patient's experience that he or she has in common with others in somewhat similar clinical dilemmas.

To focus on affect and content and to use reflective rather than interrogative techniques are usually easier to learn than to focus on the here-and-now of the evolving relationship and to respond to the command aspects of the patient's messages. These features of a person's communication can be overt and intense ("*Help me*" or "*Do something*"), or more subtle ("*The last doctor didn't help at all,*" "*If only a medicine would help,*" or "*You seem to understand so well*"). The difficulty in dealing directly with these messages is that there is nothing in most professional education to suggest that such a focus is relevant. Also, few residents have experience in personal relationships in which talking about what is going on here and now in a relationship occurs. Thus, clarifying such messages seems strange and alarming.

A particularly useful paradigm to assist beginning residents in their understanding of the moment-to-moment evolution of relationship structure is the construct of distance-regulation, which presumes that in each relationship with a future the participants must establish a mutually satisfactory metaphorical distance. The therapist must answer several questions: In psychotherapy, how close to or detached from the patient's experience does the interviewer wish to work? What distance is comfortable and productive for the patient? What are the patient's propensities?

Examples of patients who differ in their individual needs are the extreme distancing of the schizoid person, the more moderate but often tightly defended distancing of the obsessive patient, and the merger-like movements toward the interviewer of some borderline patients. Some patients make indirect requests for and movement toward closeness that is quickly undone by a variety of distancing techniques; this yo-yo maneuver is another familiar pattern. Recognizing these patterns during an interview in which one is a participant is a task of considerable complexity. I try in the seminar to introduce beginning therapists to this basic and important dimension of the psychotherapeutic process.

In particular, residents can learn to become aware of their own tendencies to retreat from certain patients or certain themes because of their own anxiety. A recent example of a beginning resident interviewing an actor-patient illuminates this process. The actor-patient has instructions to come in asking for help for a depressed, treatment-resistant wife. Old enough to be most residents' father, the actor-patient comes across as a powerful executive. Beneath the surface of his concern for and anger toward his wife, however, he is depressed, even suicidal, about a younger man having gotten the promotion he expected. He is instructed to continue his opening focus on his wife unless the interviewer assists him to share his own serious, life-threatening dilemma.

In the interview, this particular resident did not conclude that the wife was the only problem, and focused on the "patient's" feelings, helping the narrative to include the "patient's" suicidal depression. The actor-patient described having a loaded gun at his bedside table and the thoughts he had upon awakening at 4 a.m. of "*ending it all.*" At this point the dialogue went like this:

"*It is bad and you want to end it all. . . .* "

"*Yes, I want it to end.*"

"*And that's why you're here—to find a solution, to get over your depression. . . .* "

It was clear to the resident and her colleagues watching the videotape that at the very moment that the "patient" had shared his suicidal preoccupation, the resident had "bailed out" and retreated by interpreting the suicidal thoughts as a wish for a cure. We all make such errors—the important aspect is to learn from them, to go through a series of silent questions about the patient, the theme, and one's own past experience. It is not necessary to discuss them in class—although one can—but it is essential to ask one's self the proper questions. This resident appeared thunderstruck and, after a moment's silence, said, *"How, eerie—my father died when I was six and I bet that's why I didn't want to deal any more with the possibility of this man's death. He said something early in the interview . . . used an expression that reminded me . . . Well . . . this is really something!"*

In a collaborative exploration, the patient's opening comments or chief complaint are not necessarily the starting point for the interviewer's hypothesis. Rather, they often represent a cautious "feeler" on the part of the patient, often an attempt to interest the interviewer. Balint's (1972) work with general medical patients is instructive in this regard. Balint believed that medical diagnosis often is the result of a complex but subtle negotiation between physician and patient. Although there are many exceptions (recall my middle-of-the-night viral pleurisy), Balint's principle that the major focus of the interview is often negotiated applies to many collaborative interviews in psychiatry. This issue is illustrated in the following excerpts from an interview the residents view early in the seminar as an example of an in initial interview of a collaborative type.

This young woman was interviewed as part of the effort of several members of the Psychology Division at our local medical school to replicate my seminar for their graduate students. I knew nothing about the patient, including whether she was "real" or an "actor-patient." After introducing ourselves, the opening phase was as follows:

"How are you feeling?"

> *"I have bad headaches."*

"It sounds painful."

> *"They are terrible. They start in the back of my head and spread frontward. Usually they come on in the middle of the afternoon."*

"They're very painful . . . they hurt a lot."

> *"Most often they start at work and get worse on the bus going home."*

"On the bus . . . going home . . . "

> *"I've wondered about the fumes and the jostling."*

"It sounds like you've studied them."

"I really don't know what causes them . . . I hoped you could help."

In this opening, the patient sticks with a somatic presentation. Although tempted to do a "headache interview," I refrain, but my efforts to expand the focus or to help the patient reframe her dilemma are to no avail. Several minutes later in the interview, we pick up the dialogue.

"Well, let's see. You have a job you don't like, but it's the only one you can find. You feel stuck in it, and the headaches come on towards the end of your workday and worsen on the bus going home. You're very worried about what's wrong with you . . . "

"Yes . . . that's it. You see my husband is a third-year medical student and I have to work."

"He must be very busy with medical school, and you're stuck in a job you dislike."

"Even more than his being busy . . . when he's home he's got to study—usually at the library—and I can't understand what he's learning. In college I was a lit major and he a chem major . . . we studied together and talked about what we were learning, now that's gone, too."

"You've lost a big part of the relationship and, at the same time, lots is expected of you . . . to support the two of you in a job that's a downer for a lit major."

"I don't mind having to make the money . . . I just didn't expect that I'd have so little of him."

By my use of basic reflective techniques, the patient's narrative has moved beyond her headaches to include a significant loss. Several minutes later the following interchange occurred:

"You've told me you're sad and lonely. Are there other feelings?"

"No, I don't think so. . . . I don't think I'm angry."

"You're not in touch with any anger . . . "

"Well, I kind of feel that it's not justified . . . you know . . . I knew it was going to be tough and now it is . . . I should just get on with it."

"It doesn't seem right to be angry about not getting what you want."

> *"It's always been that way. Back home I was big sister . . . I was Ellen and I was supposed to take care of things. There were lots of us and my parents expected me to be a big helper . . . what I wanted wasn't . . . maybe couldn't be . . . very important. Anger, though, wasn't allowed."*

The patient has now revealed a central conflict involving her needs, her conscience, and the "real" situation she finds herself in and perhaps has helped to produce. In addition, she spontaneously reports a childhood antecedent of her current dilemma. The facilitation of the patient's narrative greatly expands our understanding. The issue of her headaches fades from a central position in the narrative, but late in the interview I attempt to place the headaches in the context of her dilemma.

"Let's see if I understand. Your husband is in medical school and you have to work and the job is distasteful. He's not available to you very much, and you've lost the closeness you had in college. You're lonely and sad . . . but if you feel angry you sweep it under the rug. All of this has some similarity to what it was like for you in your family."

"Yes. And I came thinking only about my headaches."

"What about that? Does all of this have anything to do with your headaches?"

"I really don't know."

"It is hard to know . . . "

"I do know that I don't get headaches if he is going . . . on days when he isn't going to study and we're going out to dinner . . . or be together in some way."

"That seems important . . . "

"I never thought about it before."

"Sounds like somewhere within you knew . . . "

There are many features of this videotaped interview that lend themselves to teaching about the interview process in addition to the use of reflective techniques, the facilitation of the patient's narrative, and the demonstration of a central dynamic conflict. First, attentive residents note several occasions of sighing on my part. We talk about the meaning of sighs as affective indicators, and I share with them my affective involvement in the patient's experience. Her narrative brought back into my mind the early years of my marriage during medical school. For a variety of reasons I was for the first time what my son, a generation later, called a "gunner"—that is, a front-row, take-it-all-in, lead-the-class kind of student. Terribly caught up with the excitement of medical school, my priorities did not include much time or attention to my young psychiatric-nurse, wife whose job occupied her evenings. I made little effort to consider her needs, and she was willing (at that time!) to subordinate her needs to my medical career. These memories came back to me with this patient, and my sighs were signs of the sadness I felt.

This disclosure leads the class naturally to a discussion of how one uses such experiences in the service of psychotherapy. One can be so moved by the affective sharing that an over-identification with the patient can occur, a process that sees her as victim without thinking how she may set up and participate in the marital situation that seems so painful to her. The opposite process can also occur. The therapist's affective arousal can be so painful that he or she retreats from the patient, changes the subject, or, in this instance, out of some sense of guilt is inclined to see only the patient's role in setting up a marital situation that she experiences as depriving. The possibilities are many but, at a more general level, the interview offers an opportunity to introduce beginning residents to the use of their own affective responses.

Another issue that the interview offers the opportunity to explore is illustrated in a recent seminar experience. A woman student asked if I had felt attracted to the patient. I said that I found the patient very attractive, but why did she ask? She astutely pointed out a slip-of-the-tongue I had made and had not called to the attention of the class. (I almost always do, but this bright student beat me to it.) In a particular exchange I apparently meant to say, "*Would you share with me . . . ,* " but said "*Would you share me.*" I quickly corrected my statement to its original intent, but the slip is clear (and in capital letters) on the videotape.

The implications of this slip for an evolving countertransference theme seem clear. I was not completely aware of my developing feelings until the slip, which, of course, made me attend those feelings. This slip-of-the-tongue thus gives the students and me an opportunity to discuss liking patients, feelings of a sexual nature for a patient, and a host of related subjects.

The teaching themes that evolve in the seminar include the impossibility of working psychotherapeutically with a patient unless one can find something likeable about him or her. A second theme is the inevitability of sometimes feeling attracted sexually to a patient. I suggest that, although there are some who lie, psychiatrists who work at great distance from their patients are believable when they say they never have sexual feelings for patients. Further, sexual feelings for a patient are a valuable source of information crucial to the process of psychotherapy. On the other hand, if one has sexual feelings towards most patients, the best advice is to run, not walk, to a psychotherapist.

This slip gives me the opportunity to spend an hour or so going over the ethical and legal issues involved in having sexual contact with a patient. I suggest some relevant reading, but introduce the residents to this very important subject and indicate how important it is to get consultation if one feels tempted to become actually involved with a patient.

Another issue flowing from the interview with the medical student's wife is an introduction to planning treatment. The interview itself is experienced as very illuminating by the residents—as if "*It's all there.*" I suggest that although a start has been made in understanding the patient's psychodynamics and, perhaps, suitability for psychodynamic psychotherapy, more data are needed. First I suggest that it would be important to see the husband, preferably in a conjoint interview. One might learn many things that could directly influence treatment planning. What, for example, would be the treatment implications of observing his repetitive attempts to "give" to her in such an interview—attempts that she repeatedly rebuffed? What would the implications for treatment be if one observed that the husband seemed to be a remote, unavailable person? Or what would the implications be if the couple's relationship structure was dominant-submissive or contained much conflict that the patient had not shared? Should the patient be the patient? Should the couple be the patient?

I also caution residents against proceeding with treatment without a thorough workup of the patient's headaches. That psychodynamic and organic causes can coexist is a lesson I painfully recall from the following experience.

During earlier years I had a neurotic need to know if I could hit in the big leagues. The baseball metaphor comes naturally because as an adolescent my career goal was to play major league ball. For my generation of young physicians, the big league was postgraduate training in a well known Eastern medical center. Thus, I ended up in the Department of Psychiatry at the University of Pennsylvania. One of my assignments was as consultant to the medical outpatient clinic staffed by fourth-year students and faculty. The clinic's Director, having learned of some earlier experiences of mine in internal medicine, suggested I function as both a medical attending physician and as the psychiatric consultant. In other words, I was to take my turn as students presented their patients.

The patient was a 50-year-old married lady with severe occipital headaches. She had been seen in both neurology and orthopedics, and was back for her third workup in the medical clinic. The senior student presented the history and then, once again, a completely normal physical examination. I interviewed the patient with the student present, and without much difficulty a different history was obtained.

The patient was severely obsessive-compulsive, scrubbing her floors, the woodwork of her small home, and much else repetitively because of her fears of contamination. She did all of this compulsive work despite her headaches. In fact, the only activity her headaches prevented was sleeping with and having sexual relations with her husband. She slept in an overstuffed chair in the living room. It was also not difficult to obtain the history that she was the only child of a "teenage mother gone wild" who gave her as an infant to the mother's childless aunt, herself a "religious fanatic." Throughout her childhood, adolescence, and young adulthood, her aunt had been openly anxious about the patient's sexuality, frequently cautioning her "*not to turn out like your mother.*"

The senior medical student was suitably impressed, and we decided to offer the patient a series of interviews with the student as therapist and me as supervisor. The reader must guess the ending: She never returned—perhaps because her own narrative was too frightening or, more likely, because I pushed her too hard. Six months later she was admitted with papilledema, and her occipital tumor was now evident.

The experience raises all sorts of unanswerable questions about the relationships of complex variables but, for our purposes (and despite the fact that she had no neurologic signs when we saw her), the implications are clear. Always look for biological (and social system) variables, even if the psychodynamics are "all there."

After several months of trying hard, residents begin to be comfortable with collaborative exploration. Their videotaped interviews become increasingly skillful, and it becomes clear that they can use two very different forms of interviewing as the context demands. I do caution them, however, against frequent switches back and forth during the same interview. It is apt to confuse the patient about the kind of relationship the interviewer wishes. If there are specific subjects that demand a careful directive inquiry as, for example, assessment of suicidal risk in a depressed patient, it is best to bridge the change with clear comments about what one is doing. "*We've been exploring your situation, but before finishing I must ask you a few very specific questions*" is a reasonable bridge in such situations.

Residents often ask at this stage of the seminar whether there are patients so resistant that a collaborative exploration is impossible and despite one's efforts, no narrative flow occurs. That such resistance occurs with a few patients is inevitable, and I often show a videotape consultation interview with a young man that is a clear demonstration. I saw the patient at the request of a psychiatrist taking a postgraduate seminar with me. The patient, a medical technologist, was being seen by the psychiatrist for complaints of social isolation, difficulty with fellow workers, and vague paranoid thoughts. During my interview he was remote, distant, and guarded. My attempts to use reflective techniques went nowhere. Empathic reaches were to no avail; most of the patient's responses consisted of "*Yes*" or "*No.*"

Although one interview does not prove the utter futility of continuing to try, if the patient remains as resistant as this one seems, the interviewer has learned something valuable for treatment planning. It will take a long time to effect an alliance, and a major reliance may have to be placed on treatment approaches other than psychodynamic psychotherapy.

Since I make so much of the importance of the collaborative exploration for facilitating the patient's narrative, a few words seem in order about the recent interest in narrative structure. To begin with, I suggest to residents that they need to develop a balance of gullibility and skepticism. The latter is easier for most of us than the former. Gullibility is allowing one's self to be taken in by the patient's story, and part of the work of psychotherapy is just that. One must enter the patient's story in order to develop any sense of what life is like for him or her. At other moments one must view the patient's story with deep skepticism. Situations are never simple, most persons do not understand how much all of us behave in ways that lead to the fulfillment of our inner hypotheses.

All of this leads into an introduction to constructivism, narrative truth and historical truth, and a culture's decisive impact in shaping and molding preferred narratives. These topics are introduced to the residents as intellectual background helpful in dealing with the uncertainty inherent in complex models of causality. It is not so bad, I suggest, to struggle with such uncertainty if the truth is relative. We all constantly redo our biographies, and the culture tells us what kind of stories are most valued.

Fresh from the biomedical indoctrination of eight years of education, residents are usually inclined towards an honest-to-God reality that can be measured. On the other hand, these are psychiatry residents and if they didn't have interest in complexity and the ambiguous things that make up the work of the psychotherapist, they would likely still be primarily fascinated by the magic of medicine's technology.

CHAPTER 4

Affect, Distance-Regulation, and Psychotherapy

There are many complex issues involved in understanding the role of affect in various forms of psychotherapy, and to review them all is beyond the scope of this chapter. Recently, however, Greenberg and Safran (1989) have presented such a review. They emphasize that no single school of psychotherapy has encompassed all the ways in which affect plays a role in change resulting from psychotherapy and that there is no broad integrative theory. In addition, they review the empirical literature on affect and psychotherapy, dividing it into three major categories: affect expression in catharsis, affect arousal in anxiety-reduction, and the role of affect in "experiencing" in psychotherapy.

Greenberg and Safran concluded that reviews of the process and outcome literature support the idea that high levels of patient experiencing are related to good outcome in psychotherapy, at least for some forms of psychotherapy. What is not clear is whether experiencing is a trait brought by the patient to psychotherapy or is a state arising in psychotherapy, or both. Of additional importance is the issue of defining more clearly the types of psychotherapy and the intrasession circumstances in which deeper levels of experiencing are associated with positive outcome.

The more circumscribed perspective regarding affect and psychotherapy that I wish to present involves the relationship between the patient's affect, the therapist's empathic efforts, distance-regulation in psychotherapy, and the facilitation of the patient's narrative.

During the early years the seminar was devoted almost exclusively to affect-sensitivity and the development of beginning empathic skills. The emphasis was almost entirely on the role of the therapist's sensitivity to affect in deepening the patient's exploration. The development of a theoretical perspective on distance-regulation, the role of affect in distance-regulation, and the search for the optimal working distance evolved gradually, shaped by the seminar experience itself, my psychotherapeutic work with patients, involvement in marital and family systems research, and the published work of several authors to whom I shall refer specifically later.

At the time of this writing, my perspective on the role of distance-regulation and affect in psychotherapy rests on several key constructs. At the broadest level, distance-regulation is seen as a survival-positive evolutionary process. Those individuals within a species that best demonstrated knowledge of when to flee and when to join were most apt to survive and pass on their genes. Over many generations, the biological anlage of effective distance-regulation would be increasingly represented in succeeding generations.

Thus, in any population a range of effective distance-regulation is to be found, reflective, in part, of some as yet not understood genetic mechanisms. These mechanisms may involve at least two interrelated systems, one for attachment and the other for separateness and its derivative, autonomy. That infants are born who react with either an increased tendency for attachment or for separateness is well known to mothers, and is clearly an important dimension in Thomas and Chess's (1977) work on temperament. My presumption is that these dimensions are genetically controlled, hard-wired components of the central nervous systems of newborn infants.

As Stern's (1985) work and review so aptly document, however, very early infancy is a period in which the infant is exquisitely sensitive to the surround and is capable of reasonably complex interactional sequences. Thus, the capacity for effective distance-regulation involves also the mother-child interaction and, in particular, the fit between the infant's biologic propensities and the mother's capacity for sensitivity and affective attunement.

Bowlby's (1973, 1980, 1982) monumental work on attachment and Ainsworth's pioneering systematic research (Ainsworth, Blehar, Waters, & Wall, 1978) on infant-mother attachment behaviors provide developmentalists with a major paradigm and related techniques with which to study these crucial interactions. The field has expanded greatly in recent decades, and there is little question that the identification of a number of secure and insecure infant-mother attachment behaviors is systematically possible and that such behaviors have strong predictive capacity for the child's subsequent development. The more recent extension to infant-father attachment behaviors and the search for factors that influence the attachment behaviors make this rapidly evolving field exciting.

For our purposes the benefit is that a field of systematic inquiry is centrally concerned with both separateness and attachment. The published data add much to our understanding of the early development of the individual's regulation of interpersonal distance. Whether securely attached infants will demonstrate a lifelong facility for both attachment and separateness, and insecure infants an enduring propensity either for distancing or for anxious clinging awaits the completion of longitudinal studies.

A third field of inquiry, marital and family systems research, adds additional perspectives to the development of an individual's distance-regulation propensities. Here, a system's relatively enduring interactional characteristics and their implications both for individual members and the ways in which they influence the system's interchange with its sociocultural context are emphasized. Our research team's efforts focus on identifying family characteristics that facilitate family members' capacities for both separateness and attachment (Lewis, 1979, 1989; Lewis, Beavers, Gossett & Phillips, 1976; Lewis & Looney, 1983). The separateness-attachment paradigm underlies much of current family research and is most clearly explicated in the work of Kantor and Lehr (1975).

A premise growing out of the work of many research groups is that the parental marriage establishes the template for the family's characteristic interactions that determine how much closeness and how much separateness is to be facilitated or allowed. Family clinicians and researchers alike identify interactional rules or limits to interactions leading to closeness or separateness. At a system level, dysfunction is often described in terms of the extremes of either the enmeshed family or the disengaged family. Well-functioning marital and family systems are observed to facilitate both separateness and closeness in family members.

The relevant theoretical writings and research findings are voluminous and beyond the scope of this chapter. I hope to establish that distance-regulation is a fundamental human attribute, and that a biopsychosocial perspective is potentially rewarding. Whether one emphasizes the biology, the developmental or psychological, or the social or marital-family systems perspective, distance-regulation provides a paradigm that promises linkages among them.

The three relationship structures most extensively and systematically studied—the mother-child relationship, the marital relationship, and the psychotherapeutic relationship—have a paramount interest in distance-regulation. Although they use different vocabularies and methods of study, each suggests that isomorphic processes exist and that their identification will provide important cross-fertilization with other fields of inquiry.

In the complex play of an individual's distance-regulation, affect plays a central role. Affects experienced by the individual are signals to move towards or away from another. Expressions of affects are signals to the other to move towards or away from the signaler. Affective messages may be sent to the other in a variety of ways, including words, nonlinguistic sounds, facial expressions, and body movements.

Although affective signals are often meant to encourage physical movement, frequently the signal is intended to facilitate changes in metaphorical distance, which means the relationship of one person to the experiential state of another person. When an individual begins to share the affective state of the other, he or she moves toward the other in a metaphorical sense. This sharing is often defined as empathy or attunement. If the individual notes the affective state of the other without any sense of experiencing it, the metaphorical distance remains the same. To note without experiencing can be thought of as the "objective" position. The third possibility is that upon noting the affective state of the other, the individual increases his or her metaphorical distance by, for example, changing the subject.

Each individual develops a preferred range of metaphorical distance and communicates this to the other by a variety of affective messages. When two individuals come together in a relationship with an anticipated future, a paramount issue is the need to establish a mutually acceptable metaphorical distance or range. In different language, the questions are, "*How much closeness is there to be in this relationship?*" and "*How much separateness is there to be in this relationship?*" These and related issues are most clearly delineated in systematic marital and family research, but clearly have applicability to the psychotherapeutic relationship (Lewis, 1989).

In the usually unconscious negotiation of metaphorical distance during the stage of relationship formation, the role of individual power is complex. The beginning psychotherapist must come to understand the secret of finding the preferred working distance. From the first moments of the interview, the psychotherapist is sensitive to the patient's distance-regulation messages. I believe each patient has an optimal distance—a distance at which the patient feels neither the threat of abandonment nor the threat of invasion. Finding this distance and adapting to its moment-to-moment variations facilitates the patient's experiencing and narrative flow. Some schizoid patients need initially to maintain a relatively great working distance, but for other patients their invasive propensities alert the psychotherapist to maintain considerable distance.

My experience suggests that the patient's exploration of his or her experiences is facilitated by the psychotherapist's attempts to share that which the patient is feeling. Such moments are often relatively brief intervals that punctuate longer periods during which the psychotherapist is more objective, trying to understand or planning interventions. The psychotherapist thus moves toward the patient's experience and then away from it, monitoring these different ways of being with a patient by asking himself or herself two questions: "*How long since I have felt the patient's feelings?*" and "*How do I understand the patient's dilemma?*" (Lewis, 1979a.)

Movements toward the patient's experience can be either spontaneous or deliberate. By spontaneous, I mean that one begins to feel without any sense of effort what the patient is feeling. At other times, however, one attempts deliberately to experience the patient's affect. Whether the empathic attempt is active or passive, it reflects what Havens (1986) calls the first order of clinical business—to find the other.

In psychodynamic psychotherapy, the psychotherapist's role involves two very different ways of relating to the patient's experience—one the more customary and comfortable position of some distance from the patient's experience and the other a movement toward experiencing what the patient is feeling. Beginning therapists need to learn these uses of the self early in their training. To do so involves encouraging experimentation in a safe context, "letting go" for brief periods of the distant position that characterizes so much of the professional's role. Residents can learn first to facilitate the patient's narrative with as little interference as possible. Then, as the narrative begins to unfold, they learn that both understanding and experiencing the patient's narrative are important.

Understanding, the therapist's task when he or she is at a more objective and distant position from the patient's experience, involves basically the recognition of redundancies, the repetitively occurring patterns of wishes, conflicts, defenses, affects, and relationship patterns. This type of learning is usually exciting for residents because it offers them a sense of understanding the underlying fabric of the patient's experience. It is, after all, different only in focus from their prior medical experience. As they begin to grasp the essentials, they gain a sense of beginning mastery.

Initially, it is helpful to introduce the basic concepts through the use of verbatim transcripts of exploratory interviews; later, videotapes are most effective. These early training experiences locate the action at a distance where it can be studied with considerable comfort. Much more complicated is the task of encouraging the student to work closer to the patient's narrative—to experience it as one's own.

This aspect of the process of psychotherapy must be introduced slowly. Empathy can be considered as a continuum, one end of which is the recognition of the patient's affect and sharing that recognition with the patient ("cognitive empathy") without arousing the therapist's own affect. The other end of the continuum is a deeply shared affective state ("affective empathy," "compathy," or "engagement") (Greenson, 1960; Kramer, 1989; Laughlin, 1967). Initially, one learns cognitive empathy—to develop sensitivity to the patient's affective messages and appropriate responses. The initial audiotaped and videotaped exercises described in Chapter 2 provide a relatively safe learning context.

We pay considerable attention to the language used in responding to patients' affects, and Havens (1986) describes classes of empathic statements, the simplest of which are *imitative statements.*

These are attempts to put the patient's affect into words. They offer the patient the possibility that another person understands, but do not carry the message that the therapist is actually experiencing the patient's affect. The therapist remains somewhat distant, the imitative statement is usually delivered in a bland way. If imitative statements are inaccurate, they carry the risk of suggesting to the patient what he or she ought to feel, thus, the statements can feel invasive.

Simple empathic statements (such as empathic exclamations and adjectives) represent a movement toward the patient's experience. These statements or sounds usually are not deliberate, but reflect the spontaneous response of the therapist who is intensely attuned to the patient's material.

Empathic translations are more complete reflections of the patient's affects. They express the therapist's affect openly and, in that sense, are rhetorical. In Havens's perspective, the rhetoric is in the service of the patient's uncovering and mastering the affective state. The therapist's affective expression activates the patient's affect and may offset the power of the patient's resistances. Indeed, Havens suggests that successful repression or denial of affective states may require covert external reinforcement. Empathic translations are also useful in establishing an affective baseline from which further exploration can proceed.

Residents are introduced also to Havens's description of *complex empathic statements*. These are attempts to reach the patient's more complex affective experiences, often involving several contradictory affects. He suggests, for example, that "*No one understands*" both translates a feeling of being misunderstood and suggests that I (the therapist) understand that no one understands. For beginning students, this level of empathic involvement is usually too complex to attempt, but is introduced conceptually in order to cover the variety of empathic responses.

Havens provides structure and a vocabulary for the beginner to consider in the early stages of experimenting with working close to the patient's experience. Much of this phase of the seminar, however, involves the development of sensitivity to affective messages from a relatively safe distance.

An intermediate form of moving close to the patient's experience is provided by the use of students' fantasies. In this use of fantasy, students put themselves in the patient's experience and describe what they think it might be like. This immediately raises a number of questions. Can a male resident fantasy himself a depressed woman, or a female resident an anxious man? Does age interfere? Can young adult residents enter in fantasy the experiences of an aging patient? Is it more difficult to enter the experience of a patient not conventionally attractive, or full of hate or self-pity?

These questions and others address the problem of the resident's freedom to move, however momentarily, into the strange, often frightening arena of patients' experiences. It involves an ability to set aside temporarily one's customary sense of self, to experiment voluntarily with brief periods of blurring of one's boundaries. Such loss of self can cause all sorts of vulnerabilities to appear. Most residents, however, have sufficient ego strength to experiment. Occasionally a resident simply does not participate meaningfully in the exercises. Rather, he or she pretends and, if the teacher is aware (often I am not), the pretense should be honored.

After the role playing, responding to audiotaped and videotaped material, interviews with actors, and fantasy exercises, the focus changes to careful reviews of the residents' interviews with patients. These 20-25 minute videotaped collaborative explorations provide students with an opportunity to observe themselves and others from several perspectives. From the viewpoint of the interviewer's ability to focus on and respond to the patient's affective message, much can be learned. How well does the resident note and respond to the patient's affect? Does the interviewer avoid particular affects? Such questions examine the resident's conduct of the interview.

The patient's responses to the resident's empathic efforts reflect a different perspective. An important observation is whether such efforts appear to facilitate the flow and affective intensity of the patient's narrative, or whether the patient responds with some form of resistance. The resistance may take the form of silence, a change in the subject, a brief matter-of-fact response, or any response that does not involve continuation and deepening of the narrative flow.

The residents are taught to proceed—cautiously—with their empathic efforts; if repetitive efforts lead to continuing resistance, several issues must be considered. The first is whether to confront the resistance. "*It's hard to talk about the sadness*," both focuses on and empathizes with the resistance, and is, I believe, a gentle form of confrontation, often preferable in early work with a patient.

The second issue the patient's resistances raise is the underlying reason for the resistance. Most residents, when asked, suggest that, in some way, the patient finds the affect unbearable. Although there is often validity to that generalization, the residents are introduced to the idea that what may be the source of the resistance is the patient's fear of the closeness involved in the resident's accurate empathic effort. This perspective leads to discussions about "empathic states" and fears of closeness. An empathic state is an interpersonal situation in which one person's empathic statement is accepted as accurate by the other person and results in a shared sense of closeness. Empathy may thus be considered both an individual trait and the state of an interpersonal relationship at a particular point in time.

Fears of empathic closeness are universal. It is the nature of the underlying fear, usually unconscious, that may illuminate a central strand of the individual's personality. I think that everyday, more-or-less normal fears of empathic closeness usually concern issues of self-esteem. "*If you understand my inner experience that well, will you value me?*" reflects what self-psychology calls lifelong need for self-objects. These are significant others who help maintain one's level of self-esteem.

However, fears of empathic closeness of some persons go beyond maintaining of self-esteem to involve fears of abandonment, invasion, merger, cannibalism, destruction, and death. Although few patients can report the nature of the fear directly, its intensity and particular nature may be inferred from the patient's associations and fantasies or the fragmenting, dissembling nature of the response to the empathic effort. In such circumstances, it is wise to accept, for a while, a greater working distance.

Not all the resistance to empathic closeness in psychotherapy comes from the patient. Residents, particularly early in their training, have their own fears. Many experienced therapists speculate that pursuing a psychotherapeutic career may represent an unconscious need to experience empathic closeness in a "safe" and sanctioned way.

In addition to whatever personal fears of closeness students bring to their early training experiences, there are a host of resistances brought about by changes is psychiatry. Kramer (1989) writes with eloquence about these newer impediments to the kind of listening that is central to empathic work. More than ever before, the therapist's office is crowded with intrusive insurance carriers, peer reviewers, managed health-care representatives, malpractice lawyers, and other third parties.

An even greater presence is the power of newly found interventions, particularly biologic interventions. Kramer explores the range of meanings to a therapist's thoughts about prescribing medication during the course of therapy: from a central professional and moral mandate to help, to use every means of providing relief from suffering, to a countertransferentially based need to avoid or punish the patient. The therapist's dilemma is inescapable; Kramer encourages all therapists to learn about the particulars of their feelings about various psychotropic agents. He suggests that prescribing powerful drugs may be understood as interpretations, distance-regulators, diagnostic aids, and cotherapists—as well as either the response to the mandate to help or destructive countertransference behavior. Old-fashioned psychodynamic therapy is no longer morally possible in the sense of intense, unfettered listening without attention to the indications for these new and powerful drugs.

A third group of resistances to an empathic focus is the growth and popularity of strategic interventions. For example, there are the powerful paradoxical interventions developed outside the mainstream of individual, psychodynamic psychotherapy and which are often traced to the work of Milton Erickson. Awareness of their capacity to change behavior is growing and as more students learn about them they intrude on empathic listening. Kramer points out they carry an implicit message that cleverness is a major psychotherapeutic attribute.

I share with my students the dialectic this provokes within me—a tension between two very different needs. One need is the wish to be gullible, to be taken in, to enter the patient's experience from his or her perspective. The other is the need to be skeptical and, from that disbelief, to be clever enough to intervene manipulatively and successfully in the patient's dilemma. Although the resolution of this dilemma may never be complete, careful treatment planning may guide its resolution. To know, for example, that a patient's treatment is restricted economically to 15 sessions and that the patient's psychopathology is very serious may influence one's choice of interventions.

Thus, the seminar introduces the students to a variety of resistances to empathic work, both their own and their patients'. From these interview-based discussions comes the construct of the interactional dance or ballet. Even in a single exploratory interview, a repetitive pattern of interactions may emerge involving the student's empathic movement toward the patient's experience followed by a particular response from the patient. The patient's response then provokes a particular response from the interviewer. Thus, the basic structure of the interactional ballet is comprised of a three-part interactional sequence, although longer sequences can be observed.

One of the most common sequences involves the resident's empathic movement toward the patient's experience (invited, in a sense, by the patient's expression of affect), followed by the patient's retreat (resistance of some sort), followed, in turn, by the resident's retreat, often to a more cognitive focus. This common pattern is described by Melges and Swartz (1989) as typical for borderline patients, but in my clinical and research work with couples I have seen it in a variety of circumstances, as well as across other forms of individual psychopathology. It is, for example, a common dyadic pattern in many couples in marital therapy.

The challenge is to help beginning therapists develop eyes and ears for dyadic interactions—to focus on repetitive sequences of behavior and, thus, begin to think in terms of relationship systems and their interactional characteristics. This is a leap for students whose perspective is strongly focused on the behavior of the individual. As in the introduction of other new perspectives in the seminar, it is best to start with objective material on the monitor which does not involve the student. After the perspective begins to be developed, the focus can change to the attempt to monitor the interactional ballet when one is a participant and as it is occurring. The utility of the here-and-now dyadic ballet for understanding both one's own interpersonal proclivities and the transference paradigm quickly becomes apparent at this stage of the learning process.

There are several issues needing comment about my attempts to teach the role of affect in psychotherapy to beginning students. One issue involves the general aversion within psychiatry to teaching technique in a direct manner. One basis for this aversion involves a concern that offering students such learning experiences may produce hollow technicians, robots without hearts. Kramer (1989) comments on this issue, further suggesting that students' interest in "how it is done" has in the past been considered a derivative of the child's interest in parental sex.

For me, the issue is how therapists develop "heart," and my position is that learning how "it" is done, how one attempts empathically to enter the experience of another person, is a major stimulus to the development of "heart." Although such experimentation is apt to be mechanical at first, the repetitive experience of feeling what another is feeling and then "stepping back" to understand one's own affect from the perspective of the patient's dilemma leads to the development of a broader sensitivity to the varieties of human experience. I believe this sensitivity to be "heart." One is changed by what one does.

Residents can also be encouraged to use their experiences to begin to understand patients' experiences. Most of us have experienced losses, been caught up in group contagion, failed to live up to our expectations, lost control of anger, avoided confronting another when it was appropriate—these personal experiences can be reclaimed and used to develop greater sensitivity to the particular experiences of our patients. The very specificity of such recollections assists in the development of sensitivity, rather than the resident's relying solely on the more abstract theoretical aspects of psychotherapy.

> In the attempt to encourage beginning students to use their personal experiences as a baseline for understanding their patients' dilemmas, I use examples from my life. Although most examples are common experiences, unusually dramatic episodes that throw one completely off balance psychologically are apt to help with the appreciation of major upheavals in the lives of patients. As a resident, for example, I had the following experience that introduced me to the subjective aspects of psychotic denial and its sudden disappearance.
>
> While I was driving to work one morning, my car was struck broadside by a truck running a red light. The impact knocked me out of my car and across the road. Awakening to the screams of two lady bystanders who believed me dead, I

jumped to my feet with a pervasive sense of strength and invulnerability. I directed traffic, assisted the truck driver whose arm was broken, and in general took over the situation. Despite cuts and abrasions, when the ambulance arrived I refused to enter the patient compartment and insisted on sitting with the driver. I walked into the emergency room of a familiar local hospital and before allowing an examination to proceed, insisted on calling my wife.

When I heard her voice, I fainted. Awakening some moments later on the gurney with doctors and nurses around me. I promptly lost all sphincter control and experienced the most searing, profound anxiety of my life. Trembling, crying, completely out of control, I gave myself over to the care of the emergency room team.

Although there is much about this brush with death that illuminates various aspects of my personality, for me personally the episode introduced me to what it is like to experience severe, overwhelming, fragmenting anxiety, to appreciate the miracle of psychotic denial, and to have that miracle suddenly disappear. In seeing patients with severe panic or patients whose denial for whatever reasons is of the magnitude that mine was on that day, I have the conviction that I have been there, and having been there I am less afraid to approach the patient's experience.

All therapists use their personal experience in this way. What I try to do in the seminar is to introduce the concept early in the students' training, and to keep in front of them the question, "*If you believe you understand the patient's experience, what is there in your life that is similar or in some way comparable?*" Retrieving such personal experiences may assist one in entering the experience of patients.

A few comments are in order about my approach to theory in the seminar. Some of my colleagues have asked why I don't teach a particular, well-circumscribed theory. There are many reasons why I don't do so. Most of all, I suspect, it is because my own theory (or theories) keep evolving and I have to keep teaching and writing to know what I think. Although I like to think of myself as a pluralist—a clinician able to use different theoretical approaches in accord with each patient's particular needs (and support that desire for pluralism by treating individuals, couples, and families)—in my work with individuals I am clearly eclectic. I borrow heavily from psychoanalytic theory, self-psychology, interpersonal theory, and existentialism. I simply cannot feel intellectually comfortable attempting to fit all I have learned into one theoretical perspective.

Havens's (1973) observations about the four basic schools of psychiatry (objective-descriptive, psychoanalytic, interpersonal, and existential) brought a welcome sense of order to my mind when I needed it. His call for pluralism at the level of treating individuals led me to try—and although objective-descriptive psychiatry with its emphasis on diagnosis and the psychoanalytic psychiatry of my earlier years could be integrated, interpersonal approaches (to the individual) and purely existential approaches seemed beyond my reach much of the time.

During my professional life there have been changes in psychiatry's primary perspective on psychopathology. It has moved from an intrapsychic focus (objective-descriptive and psychoanalytic perspectives) to become more interpersonal and existential. I understand the major impact of self-psychology with its focus on affect and empathy to reflect a more existential orientation than was apparent in early psychoanalysis. The concept of the self-object joins psychoanalysis and interpersonal psychiatry. The major development of marital therapy and family therapy reflects an increased reliance on interpersonal perspectives. Research-based efforts to develop reliable approaches to psychodynamic formulations often focus on interpersonal episodes and central relationship conflicts, thus also reflecting a greater emphasis on the interpersonal.

In the therapist's development of a psychotherapeutic self many therapists go through phases (Havens, 1973; Lewis, 1978; Ralph, 1980). Many therapists start with an objective-descriptive orientation, then move to a more psychoanalytic perspective. Some remain more or less pure in their psychoanalytic orientation, but others go on to incorporate emphases on interpersonal psychiatry and, perhaps later in life, a more existential orientation. I can see evidence for this in myself and some colleagues.

There is also the issue of to what extent a therapist's therapeutic perspective is reflected in his or her work. Audiotapes and verbatim transcripts often suggest that many therapists are more eclectic than they acknowledge.

I hope to introduce the students to multiple theoretical perspectives, suggesting that each has its strengths and limitations, ways of conceptualizing psychopathology, definitions of the preferred working distance, and preferred interventions and language. It is sometimes useful to take a single patient's expression and conjecture about the therapist's response if he or she were an advocate of one of the four major schools.

Although I place heavy theoretical emphasis on the role of affects in psychotherapy and believe that empathy is a major therapeutic intervention, there is much more to psychotherapy. It is to these equally important aspects of teaching and learning psychotherapy that I now turn.

CHAPTER 5

Relationship Structure

For beginning students, understanding the state-trait distinction as it applies to the interview involves a giant leap in complexity. They are asked to observe interviews from the perspective of a two-person interaction and to consider each person's interview behavior as a reflection of the state of the ongoing interaction. This involves temporarily putting aside their traditional focus with its customary assumption that the patient's behavior reflects more-or-less enduring traits.

An early exercise in the seminar introduces this shift in perspective. The students are asked to comment on short interview segments such as the following:

"You didn't feel anything when you had to repeat the course?" (matter of factly)

"No—it was just one of those things." (without affect)

"Nothing at all"? (blandly)

"Nothing." (without affect)

"Can you feel anything now?" (remains bland)

"No."

The students invariably focus on the patient's lack of affect and discuss it from the viewpoint of defense against painful affect. Although I acknowledge that such an observation is a reasonable perspective, I ask the students to try another observational stance: to view the segment interactionally and to describe what they see and hear. From this perspective it is quickly apparent that both interviewer and patient contribute to the absence of affect. The interviewer is as bland as the patient; it is as if there is a collusion of sorts to keep the interview free of affect.

It is simply not possible to tell on the basis of these data whether the patient's lack of affect is defense—perhaps even a trait—or whether the patient is responding to the interviewer's lack of affect and that what is observed is a process involving mutual cuing. Indeed, to complicate the issue further, it is possible, even likely, that there is validity to each perspective, that the patient is both defending against painful affect and responding to the interviewer's lead. One's perspective or theoretical orientation influences directly what one observes and concludes.

The differences between these two perspectives are profound: The patient-only focus emphasizes his or her separateness, whereas the interactional focus is fundamentally contextual. There are many reasons that make understandable the difficulty most students have when asked to let go temporarily of a perspective that focuses on the individual person and his or her acts and to observe and understand the individual as a participant in interactions that shape his or her behavior. In his moving essay, *The Tree*, John Fowles (1983) discusses Western consciousness as directing our attention to individual organisms and acts. He describes his reaction to the garden of the great classifier of botany, Carolus Linnaeus, as a visit to the site of an intellectual atomic explosion, a turning point in the way we think about the world. Linnaeus studied individual specimens out of their interactional contexts, thus directing our attention to the separate individual (even the tree) as freestanding and without a context that modifies structure and function.

If to this preoccupation of Western consciousness with individuals and their boundaries is added the intense emphasis of science, medicine, and psychiatry on studying and treating individuals separate from their interpersonal contexts, the teacher introducing this complementary perspective to beginning students is confronted by a considerable task.

It is useful to cite a variety of methods in the attempt to introduce students to the advantages of having multiple perspectives from which to understand and intervene with patients and their interpersonal systems. One method is the case presentation in which the application of a different perspective results in dramatic change.

Mrs. A,* an aging beauty with a finely developed ability to portray her desperation in dramatic terms, was a long-term psychotherapy patient of a senior colleague. She drank excessively, misused prescription drugs, and was often suicidal. Her life seemed organized around a pervasive depression behind which one often glimpsed homicidal rage. She was hospitalized often and sometimes for intervals of many months.

Her husband, a successful and influential businessman with rural roots, was stoic in his response to his wife's tumultuous clinical course. A man of few words without evident interest in or knowledge of psychological constructs, he persisted in distancing himself from his wife's treatment with a clear attitude of "*Something is wrong with her—she needs fixing.*"

Their relationship began at the state university where he was a poor-boy football hero on a scholarship and she was the sorority beauty queen from an affluent family. They married after graduation and quickly settled into a conflicted relationship in which she felt emotionally deprived by his affective remoteness and he felt nagged and retreated even more to his growing business.

After their several children were of school age, Mrs. A began to drink excessively and to exhibit the symptoms of a major depression. Following a nearly fatal suicidal attempt, she was hospitalized. In contemporary diagnostic terms, she was seen as a person with a borderline personality disorder, a major depressive disorder, and alcoholism.

Mrs. A's treatment program never went well for very long. In the hospital, she was well controlled and, on the surface, cooperative. When she left the hospital to be seen as an outpatient in intensive psychodynamically oriented psychotherapy, she quickly regressed. A variety of antianxiety agents, antidepressants, and major tranquilizers were used with no sustained effects. Because of her sensitivity to any hint of rejection, an MAO inhibitor was used without clinical improvement. I first saw Mrs. A in consultation when she developed a delusional transference to her therapist.

*This clinical vignette adapted originally from one for publication in *Cures by psychotherapy: What effects change?* (Myers, 1984). Reprinted with permission.

Mr. A steadfastly refused to be involved in any form of therapy. After a particularly frightening suicidal attempt by his wife, he reluctantly agreed to see me for 10 exploratory interviews. In these interviews, he described the economic and emotional deprivations of his early life without, however, any apparent recourse to the affective components of those memories. He was an emotionally remote man genuinely perplexed by his wife's behavior. He valued rugged individualism, loyalty, and "gutting it out." After the requisite 10 interviews, he bid me a matter-of-fact farewell—he had fulfilled his contract.

Mrs. A's treatment continued for years and ultimately she broke it off when her therapist insisted that she give up alcohol, attend AA, and stop abusing prescription drugs. His heroic confrontation (not the first) led this time, however, to her rageful retreat.

I heard no more about either Mr. or Mrs. A until one day several years later when he called crying and asked to see me. I insisted that Mrs. A accompany him and saw them together the next day.

He had pulled himself together and focused on her continuing symptoms. I insisted that we talk about his request for help. After some time, tears ran down his cheeks and he said that he had become increasingly aware that he was never going to be truly close to another person and that this realization saddened him. Mrs. A watched disbelievingly. I asked her to hold his hand, which she did.

My understanding of their marital dynamics involved a picture familiar to all therapists who see couples. Although different descriptive labels are used, the relationship structure is understood as chronically conflicted, with both spouses having more than the usual underlying fears of closeness and the need to maintain distance. Chronic conflict is thus seen as a way of relating that involves a tenuous connection without very much in the way of closeness (let alone intimacy) that might impinge upon their shared underlying fears. Although Mrs. A was the pursuer and Mr. A the pursued, it is important to understand that each feared closeness. From another perspective, this couple reflected the familiar theme of hyperadequate husband-inadequate wife. What is central to this perspective involves a mutual projection system in which all or most of Mrs. A's sense of competence is projected to her husband, and all or most of Mr. A's sense of incompetence is projected to her. Each accepts the other's projections and behaves in ways that confirm the validity of the sharp distinctions of level of adequacy in this marital system.

Thus, I understood Mr. A's depressive and vegetative symptoms as a remarkable opening in what had been a rigidly patterned marital system.*

On the basis of my formulation of the marital dynamics I asked each spouse to request one thing they really wanted from the other. She requested to be held 30 seconds each day. He agreed, although asking for numerous clarifications regarding the details. He asked that she stop drinking, a request I ruled unreasonable. He then asked that she stop bringing the portable television set to breakfast, their only meal together. She agreed, and they went off to return the next week.

I saw them together five times using behavioral interventions designed to increase slightly the level of closeness in their relationship. The last intervention involved structural exercises designed to re-establish an element of physical affection in their relationship. His depression, anorexia, and insomnia had cleared by the third marital session. Six months later I received a letter of gratitude from them, and several years later they called asking me to arrange treatment for a depressed adult child of theirs. At that time I inquired about how they were doing. They were happier than ever before, but for purposes of this description it is important to note that following the five marital interviews Mrs. A had stopped drinking, had used no prescription medications, and had not experienced any further depressive symptoms.

Single case clinical reports often raise more questions than they answer, and my experiences with Mr. and Mrs. A are no exception. I use this case report, however, with beginning residents both to illustrate the utility of using different perspectives and to empha-

*In another context and in response to a survey indicating that over 50 percent of psychiatrists feel that it is either unethical or malpractice not to prescribe antidepressants in such situations, I argued that it could be considered unethical to prescribe such medications in this clinical situation.

size the power of relationship structures to mold the behavior of the participants. Mr. and Mrs. A were locked into a rigid and dysfunctional relationship structure that for years had shaped important parameters of their behaviors, including Mrs. A's long-standing psychiatric syndromes. The issue is not the superiority of one perspective, for other case reports could as easily demonstrate the dramatic advantage of a switch to a biological or a psychodynamic perspective.

When one is introducing students to the concept of relationship structure, both case reports and segments of interviews are valuable because the interaction is "out there" in the sense that the students are not participants. Even these materials, however, result in considerable confusion, in part because beginning students have no theoretical framework on which to base their observations: They do not know what to look for. They need the fundamentals of a theory concerning relationship structure, a theory that directs their attention to a small group of interactional variables useful in assessing relationship structure.

When a theoretical perspective has been established and they attain some facility at recognizing relationship structures that are "out there" in material not involving themselves, the students are encouraged to apply this perspective in observing and reacting to videotapes of their own interviews. Even in these situations, however, their videotaped interviews are "out there" in time, the tape can be stopped for discussions, and the students become increasingly aware of how little awareness they had of their own interactional behavior as it occurred.

Although most students become more knowledgeable about their interactional tendencies—many of which, not surprisingly, can be understood as essentially distancing in nature—I suggest to them that increasing sensitivity to the relationship structures they are establishing with different patients, *as they are establishing them*, is a lifelong pursuit. To the extent that a therapist can observe herself or himself in the process, the greater the ability both to understand the patient's relationship propensities and to avoid the development of relatively rigid and destructive transference and countertransference fixations.

The presentation to students of the theoretical aspects of relationship structures involves a number of decisions. I have selected the three two-person relationship systems that have been most extensively studied by direct and systematic observational methods; the marital relationship, the infant-mother relationship, and the psychotherapeutic relationship itself, particularly that small body of work on the patient-therapist alliance. I emphasize the marital relationship both because of the relatively large body of empirical research and because of my direct experience in this area. My focus on the competence of relationship structures comes out of this research. I expose the students to my search for parallels between marital relationship structures demonstrated empirically to be highly competent and secure infant-mother attachment systems and effective psychotherapeutic alliances. Self-report studies seem less valuable to me in regard to relationship structure because of the difficulty most people have in correctly appraising their roles in shaping the nature of their important relationship structures.

The concept of relationship structure originated in the past three decades of research in marital and family systems. This research has taken as its primary focus marital and family relationships as systems—that is, entities greater than the sum of their parts that develop characteristics not completely predictable from the characteristics of the parts themselves. The *structure* of the system refers to the relatively enduring patterns of repetitive interactions. Relatively enduring does not mean rigidly fixed; rather, it means capable of change when faced with a significantly altered context. Marital system researchers directly study the patterns of interactions occurring between the spouses.

I suggested in the preceding chapter that a useful paradigm with which to understand the pattern of marital interactions is that of distance regulation. As two persons come together in the early stages of a relationship that has a future, they must somehow decide how much closeness and how much separateness will be involved in their relationship. This search for an agreeable balance of closeness and separateness is a central task of the stage of relationship formation, a balance seldom dealt with directly, but its subtle negotiation is the metacontent of much that transacts between the individuals.

Within the question of "how much" closeness and separateness are, of course, a number of issues involving who decides, what happens in the case of disagreements, how often and intense such disagreements are, what forms of closeness and separateness are sanctioned or prohibited, and others.

It is, then, the problems surrounding relationship formation that are central to the concerns of this chapter. I will rely primarily on our research and summarize it briefly. The interested reader can pursue these issues in greater detail in recently published accounts of our longitudinal study of young families (Lewis, 1988, 1988a, 1989; Lewis, Owen, & Cox, 1988).

There are, to my knowledge, no direct observational studies of marital relationship formation, for to accomplish such would involve studying relationship systems from their outset. Our own work begins, therefore, with several assumptions. The first assumption is that studying marital interactions early in the history of the relationship (during the couple's first pregnancy) informs as to the relationship structure they have evolved prior to their pregnancy. We identified five variables that we believe in concert are involved in the establishment of the relationship structure. These variables are closeness, intimacy, commitment, power, and separateness-autonomy.

Closeness, intimacy, and commitment represent different aspects of the degree of connectedness the system will manifest. By closeness is meant the amount or level of sharing of values, interests, activities, friendships, and satisfaction with sexuality. Intimacy is defined as the system's ability to facilitate the sharing of vulnerabilities, those very private feelings, hopes, dreams, and fears. Commitment refers to both the intensity and the loyalty to the relationship in comparison to other relationships in which the participants are involved.

Separateness-autonomy is defined as the system's capacity to facilitate each participant's involvement outside the relationship. Power, or more specifically, the distribution of power, is a judgment of how the system deals with the requirement of arriving at guidelines about who decides what, how disagreements are resolved, and, eventually, how the decisions about the balance of closeness and separateness are determined. When seen from this perspective, the negotiation of power distribution is central to the crystallization of a more-or-less enduring relationship structure.

Both our current study of young families and two earlier descriptive studies of families containing adolescent children suggest that at a macroscopic level there are but a small number of basic marital relationship structures, and that these structures can be identified reliably from relatively brief segments of videotaped marital interaction (Lewis, 1989; Lewis, Beavers, Gossett, & Phillips, 1976; Lewis & Looney, 1983). We also present evidence that the marital relationship structures have been noted in the research of other investigators, and that it is possible to order these structures along a continuum of relationship competence (Lewis, 1989). The continuum of competence is based on value judgments regarding the purposes of a marital relationship. Our work is based on the idea that the marital relationship has two cardinal purposes: the facilitation of the spouses' psychological health and the facilitation of the psychological health of any children born to the couple. These purposes are considered to be primary only in those contexts in which physical survival is relatively assured.

Marital systems that best accomplish the two cardinal purposes demonstrate high levels of closeness, intimacy, and commitment associated with high levels of separateness-autonomy. Power is shared relatively equally in such marital systems.

Intermediate between the highly competent marital systems and relationship structures seen as dysfunctional are competent-but-pained marital systems. These relationship structures are competent from the perspective that they facilitate the healthy development of adolescent children, but are pained because of an underlying marital conflict regarding closeness, intimacy, and commitment. One spouse is chronically dissatisfied with these aspects of the relationship, and the relationship structure does not promote his or her psychological health. The marital system is thus seen as having a conflict about closeness and intimacy and containing relatively low levels of commitment. Separateness-autonomy is usually at a high level in both spouses. The distribution of power is most often skewed, with the distant spouse being more powerful.

Next on the continuum is the dominant-submissive marital relationship. Although many such relationship structures "work," they are seen as dysfunctional because of an increased probability that either the submissive spouse or one of the children will develop psychiatric symptoms. This relationship structure, for example, may be found in the marriages of depressed women or in the parental relationships of children with behavior disorders.

Closeness and commitment vary; intimacy is absent. The autonomy of the submissive spouse is often compromised, and the rigidly skewed distribution of power is the central relationship characteristic.

Clearly dysfunctional is the chronically conflicted relationship structure. This type of system is understood as the inability of the participants to negotiate successfully any semblance of agreement on the basic parameters of their relationship. Often understood as reflecting the tenuous connectedness of spouses who share underlying fears of closeness and intimacy, such relationships often endure for the lifetimes of the participants. They are, however, relationship structures that do not facilitate the psychological health of either the spouse or their children. Low levels of closeness, the absence of intimacy, and little commitment are associated with both considerable separateness-autonomy and the characteristic chronic power struggle.

Severely dysfunctional marital systems are of two types: the symbiotic and the alienated. The symbiotic relationship structure involves the absence of separateness-autonomy and the blurring of the individuals into a "we-ness" that has a primitive quality. Alienated relationships represent the opposite structure—very little meaningful (or no) connectedness associated with high levels of separateness-autonomy. Each of these relationship structures is associated with the probability of high levels of psychopathology in the parents and their children; neither nourishes healthy development.

Thus, this approach to the study of relationship structures suggests but a small number of types that in terms of competence can be understood as ranging from structures that are flexible and collaborative to those which are rigidly patterned in either a dominant-submissive or chronically conflicted configuration to those that are either merged or alienated.

As indicated in the preceding chapter, the systematic study of infant-mother attachment behaviors started with the theoretical contributions of Bowlby (1973, 1980, 1982) on the concepts of separation and attachment. Out of his work has grown a huge developmental research literature, much of it using Ainsworth's empirically based systematic approach to the study of the infant-mother interaction (Ainsworth, Blehar, Waters, & Wall, 1978).

Ainsworth's research paradigm involves the Strange Situation Task in which the one-year-old infant and his or her mother are exposed to a series of programmed separations and reunions in a standard setting. The outcome measures are derived from the child's behavior and include measurement of the child's freedom to explore standard toys in the research playroom, the child's response to the entrance of a stranger, the intensity of the child's protest when left by the mother, and the child's manner of dealing with the mother's return.

A small number of typical attachment behaviors have been identified and replicated in many studies, and these attachment behaviors are understood not to be aspects of the child's temperament because the type of attachment may differ with mother and father. Rather, the attachment behaviors are interpreted to be the child's manifestations of the nature and intensity of the relationship bond with the parent.

A secure attachment is manifested by the child's freedom to explore, significant protest upon the separation from the mother, and a reunion that involves moving to the mother without inconsolable clinging. Two types of insecure attachments have been identified: anxious and avoidant.

In the anxious attachment, the child is less free to explore, protests the mother's departure significantly, and reunites with the mother in an often inconsolable, clinging way.

Avoidant attachments are characterized by freedom to explore, less protest upon the mother's departure, and little, if any, movement toward the mother upon reunion. Although there are subclasses of each of the basic types (and perhaps a fourth basic type—disorganized), for our purposes it is important to emphasize that the infant-mother attachment behaviors measured when the infant is one year old have proven to have substantial predictive strength in terms of important parameters of the child's subsequent development. It has also been possible in our own work to establish these types of attachment behaviors along a security of attachment continuum that has proven also to have significant predictive capacity (Lewis, 1989; Owen & Henderson, 1989).

For the purposes of this chapter, however, I wish only to emphasize that secure attachment behaviors involve an intense bonding that serves as a secure base from which the child is able to explore the immediate context. Both connection and separateness can be understood as qualities underlying the secure attachment.

The anxious attachment involves less capacity for exploration or separateness and what can be understood as an overly dependent, somewhat angry type of connection. The avoidant attachment can be understood as a very different and, in some ways, less intense form of bonding or connectedness associated with the capacity for separateness.

It is possible to understand at a very general level some similarities between these infant-mother attachment systems and several of the marital systems described earlier. In particular, the secure attachment behavior shares with the highly competent marital systems a facilitation of high levels of both connectedness and separateness. The parallels involving anxious and avoidant attachments and dysfunctional marital systems are also noteworthy. In these dysfunctional systems, as with insecure attachments, it appears that either the capacity for separateness or the capacity for connectedness is compromised.

There are no studies, however, that directly compare relationship competence across these two different relationship systems, using direct observational methods. Although our study of young families demonstrates that spouses with dysfunctional marital relationships before the baby's birth are more likely to be subsequently involved in insecure attachments with their one-year-old children, this finding does not directly answer whether the relationship structures (marital and parent-child) are comparable (Lewis, 1989). The only study exploring the relationship between romantic love and attachment behaviors was designed to illuminate the continuity in individuals between self-reports of early relationships with parents and adult romantic relationships (Hazan & Shaver, 1987). Although the findings were generally positive regarding the matter of continuity, the issue of similarity of structure can only be inferred.

When the focus is turned to the patient-therapist alliance, the first issue needing clarification is the type of psychotherapy. In his descriptions of the four major schools of psychiatry, Havens (1973) suggests that the four schools differ along all major parameters of the treatment situation, including the preferred nature of the alliance. Although respect for the patient transcends theoretical orientation, each school, for example, calls, at least implicitly, for different distributions of power within the structure of the patient-therapist relationship.

Empirical studies of the psychotherapy relationship have not, to my knowledge, used the insights and methods from systematic approaches to the structure of either the marital or mother-child relationship. Luborsky and his colleagues in the Penn Psychotherapy Project made important research contributions to the complexities involved in understanding how psychotherapy works, but from the perspective of this chapter I wish to focus only on their research on the helping or therapeutic alliance (Luborsky, 1976; Luborsky, Crits-Christoph, Mintz, & Auerbach, 1988; Luborsky, McLellan, Woody, O'Brien, & Auerbach, 1985).

They demonstrate that the nature of the alliance is an important predictor of successful psychotherapy. Two types of alliances are defined. Type 1 is characterized by the patient's experiencing the therapist as supportive and helpful, with the patient as recipient. Type 2 is based on shared responsibility, a "we-ness" in which collaboration is the underlying process. In a 1976 contribution, Luborsky suggests that the Type 2 alliance is a more substantial achievement than the Type 1 alliance, and can be of more lasting value to the patient. Twelve years later the findings of the Penn Project were summarized, and those regarding the helping alliance are of importance to our interests (Luborsky, 1976).

Using three methods of measuring the helping alliance (two of which are observer-based and one of which is based on self-reports from patient and therapist), important findings were noted. The first involved the fact that all three measures of the alliance taken during early interviews were correlated with successful psychotherapy. The second involved the high correlation between Type 1 and Type 2 alliances. I shall return to this finding later as it seems crucial. The third important finding was that the Type 1 alliance was more frequently observed than the Type 2 alliance and, although the Type 2 alliance was predicted as more likely to become evident in later sessions, this increase was not statistically significant.

Thus, the Penn Project provides strong support for the importance of the helping alliance in the outcome of psychotherapy. From the perspective of relationship structure, however, we are left with tantalizing uncertainty. Luborsky's 1976 distinctions of two types of alliances rested directly on the distribution of power in the patient-therapist relationship. Type 1 involves a powerful therapist who gives help to a less powerful "recipient" patient, while Type 2 is a collaborative relationship with greater sharing of power.

However, identifying the two distinctly different relationship structures is clouded by reliance upon patient statements about the general issue of being helped by the treatment, optimism about its outcome, and feeling a rapport with the therapist. Also, the very high correlation between the two types of alliances indicates the failure of the measures to distinguish the difference between the two. The more frequent finding of Type 1 alliances is, I believe, related to the definitional inclusion of the broad and nondistinguishing parameters noted above.

Thus, despite the major contribution of this project to our understanding of helping alliances and treatment outcome, the study does not provide all that one might wish in regard to the issue of relationship structure in the psychotherapeutic relationship and its connection, if any, to the relationship structures found in well-functioning marital systems and secure infant-mother attachments.

This brief review of selected aspects of the empirical evidence for relationship structures that facilitate individual health in three different types of relationships is, at most, suggestive. It suggests that studies aimed at demonstrating that relationship structures that facilitate high levels of connectedness and high levels of separateness-autonomy may in three very different types of relationships (marital, mother-child, and psychotherapeutic) facilitate the growth of the participants and, as a part of that growth, their abilities to explore and expand their worlds.

The possibility of a parallel between well-functioning marital-family systems and effective psychotherapy has been a signal theoretical contribution of Beavers (1977). A major member of the Timberlawn Psychiatric Research Foundation research team at the time of the earliest studies of family health, Beavers went on to assess a variety of forms of psychotherapy from the viewpoint of family characteristics demonstrated to facilitate healthy psychological development in family members. The eight family system characteristics used in his analyses include:

1. *A Systems Orientation.* Beavers focuses on the assumptions of family members in well-functioning families that (a) people

must have a viable ongoing social system for basic needs to be met; (b) causes and effects are interchangeable; (c) there are multiple, rather than single, causal factors; and (d) people are finite and limited.

2. *Boundary Issues*. The well-functioning family is characterized by clear individual boundaries without distancing.
3. *Contextual Clarity*. A major aspect of the well-functioning family's clear context involves generational boundaries. In such families there is no blurring of authority and the strong parental alliance does not permit the formation of dysfunctional family triangles.
4. *Power Issues*. Beavers focuses on the consistent finding that the parents in well-functioning families share overt power; members' roles are complementary rather than symmetrically competitive.
5. *Encouragement of Autonomy*. Well-functioning family systems encourage individual autonomy by facilitating the ability of each family member to take personal responsibility for his or her thoughts, feelings, and behavior, to be open and clear in communications, and to respect others' expressions and individual subjective reality.
6. *Affective Issues*. The emphasis here is on well-functioning families' warmth, optimism, and encouragement of open expression of affects. Empathy is high and conflicts are usually quickly resolved.
7. *Negotiation and Task Performance*. Beavers emphasizes the capacity of optimal families to raise children well, to negotiate differences, to integrate ideas from all family members, and to solve problems effectively.
8. *Transcendent Values*. The focus here is on a family's belief system that allows members to make sense of events, to accept aging, loss, and death, and to maintain hope.

Using these documented characteristics of well-functioning families, Beavers assesses the following types of psychotherapy: psychoanalytic, interpersonal, existential, gestalt, client-centered, and behavioral. His analyses describe the ways each modality both incorporates and excludes different growth-facilitating characteristics of healthy systems, and he suggests that gradually a single form of a "synthetic, pluralistic, scientific psychotherapy" (p. 250) will evolve that will eliminate the closed-system aspects contained in each of the six current forms of psychotherapy.

Beavers provides a clear conceptual linkage between highly competent marital and family systems and the psychotherapeutic alliance. More than any other theoretician, he has suggested that relationship structures demonstrated to be growth-facilitating in one type of relationship have direct applicability in another type of relationship.

Most beginning students have difficulty in adopting this perspective, and I wonder if the difficulty might have something to do with giving up the taking of sides.

For many, taking sides is a deeply embedded interactional pattern going back, I suspect, to early experiences within our families of origin. As a little boy, I recall clearly taking my mother's side whenever her conflicts with my father surfaced. It seemed clear to me then—as it does not now—that she was invariably right, the injured party, or whatever, and my father was invariably wrong and in some ways the victimizer.

I can recall also a similar reaction to disagreements between two friends. As a child and through early adolescence, I found that one friend was clearly "right" and the other friend clearly "wrong." As with conflicts at home, it seemed natural to take sides, and I can't recall any doubt about the rightness of my choices.

Later, taking sides was also evident as part of my adolescent responses to social issues. I unambiguously took one side, whatever the issue.

In coming to grips with the idea of relationship structure, one has to give up a good deal of the taking of sides. Inherent in the concept is the idea that relationship structures are invariably negotiated. Both parties are participants and the behaviors of each are, at the same time, both cause and effect. Appreciation of this joint participation in the elaboration of relationship structure in enduring relationships makes side-taking, with its associated sense of certainty, much more difficult. As we understand better the complexities unveiled by this perspective, we must find ways to be comfortable with uncertainty and ambiguity.

All of this was brought home again earlier on the day of this writing as I was seeing a 30-year-old patient with a schizophrenic-like disturbance. We have worked together for 12 years, and the patient had done relatively well until a year ago. Following several losses, she withdrew from social contacts and began experiencing intrusive voices arguing over her soul.

She has experienced both parents, but particularly her father, as very intrusive. Her parents handle her finances, tell her what to wear, approve or disapprove of her friends, and in many ways appear to be overinvolved in her life. It is easy to understand the patient as victimized by parents who share the trait of intrusiveness.

If one changes the perspective, however, and tries to understand the patient's parents' intrusiveness as responsive to certain of her behaviors, a much greater appreciation of the complexity of their relationship structure is possible. Thus, her passivity, failure to deposit checks, and failure to take responsibility for herself in other ways can be understood both as a result and a cause of her parents' intrusiveness.

The advantage of not taking sides are numerous, but among the foremost is the idea that by not taking sides one sees both or all participants as sharing responsibility for the nature of their relationship. It is, however, understandable that beginning students cling to the older certainties.

A second issue that comes to mind is the need to emphasize that in the seminar the students are introduced to the concept of relationship structure, but clearly it is only an introduction.

The interpersonal perspective of which relationship structure is a derivative is a dramatic shift in the way of observing and knowing the world: a shift away from the perspective of a fixed reality in the form of diagnoses and diseases; a shift toward a realization that one is always not only involved in, but also shaping the nature of one's intercourse with others. I teach this perspective, however, not as the "truth," but as another perspective, another prism through which to view and understand. That such a perspective often has powerful therapeutic utility is, I hope, made clear to my students.

CHAPTER 6

An Introduction to the Cognitive Work of Psychotherapy

The psychoanalytic parade started after World War II, the bandwagon was ridden by psychiatrists returning to civilian life who brought with them an introduction to psychodynamics and psychoanalytic psychotherapy. An era of biological psychiatry with its electroshock, metrozal, and insulin coma came to an end.

There was great enthusiasm for the new bandwagon of psychoanalytic psychiatry. Training programs organized their curricula around psychodynamics and psychotherapy, and many of medicine's finest young students opted for careers in the new psychiatry.

The appeal was primarily intellectual; psychoanalytic psychiatry was first and foremost a new system of understanding. Young physicians interested in *how* and *why* rather than *what* were most likely to be attracted. Diagnosis, the "what" of psychiatry, quickly fell into disfavor as the dynamics of conflict and developmental arrest flourished.

Insight became the password of the new psychiatry. Truth frees one from neurotic bondage; psychotherapy was most often experienced as an intellectual search for the truth. When truth failed to cure it was divided into two forms: intellectual insight and emotional insight. The former was generally ineffective, the latter almost always productive.

Thus, the first postwar generation of psychotherapists was trained at the three-tiered cognitive altar of confrontation, clarification, and interpretation. Little attention was paid to exactly how one went about these interventions. Audiotaping psychotherapy interviews was forbidden, and there were no films upon which the art of the masters was recorded. It was no surprise, therefore, that when I listened to audiotapes of stalemated psychotherapy (my own and those of supervisees) the most common finding was that of a relatively sterile, intellectual dialogue. Further, it is no surprise that the initial phase of my seminar was concerned solely with feelings, affect sensitivity, and empathy. After I had taught the seminar in that way for several years, it seemed reasonable to add a cognitive component, to find ways to introduce beginning residents to how the more cognitive aspects of psychotherapy are actually done.

As briefly noted in Chapter 2, I focused these introductory exercises on three groups of processes: those that increase narrative flow, those required in making the clinical formulation, and three classical cognitive interventions; clarification, confrontation, and interpretation.

These more cognitive aspects of psychotherapy have always been welcomed by residents. Learning about them and how to go about actually doing them do not have the same sense of unfamiliarity—even danger—that exercises centering on affect, empathy, and fantasy seem to contain. They are more like what one does in general medicine and there is a familiar sense of "objectivity." One is dealing with psychopathology, a construct that can be made to seem of a different order than the patient as person. The interpersonal distance is comfortably like that in any professional relationship—"it," the psychopathology, is "out there." There is no mandate to experience what the patient is experiencing, no request to "let go" and enter the patient's world, no injunction to cast aside preconceptions and be taken in. Indeed, a skeptical stance, particularly about the patient's casual explanations, is welcomed.

The easiest place to start is with the relatively simple ways of facilitating narrative flow. "*How easy it has been to talk with you*" or "*I can't believe all I've told you—even some things I didn't know I knew*" often punctuate the end stages of interviews in which a patient's story has been skillfully encouraged.

To encourage storytelling, to stay out of the patient's narrative, not to interrogate the patient—this is the first and, in some ways, most fundamental task. For beginners, it is an act of faith, faith that the patient will tell you what you need to know, faith in the act of listening without relying on asking questions.

Questions are the natural enemy of narrative construction. There is, of course, nothing wrong with an occasional question. However, interviews dominated by the interviewer's questions interfere with the flow of the patient's story, suggest what the interviewer wishes to hear, and mold the relationship into a dominant-submissive structure. As students of marital and family systems have learned, asking questions is a powerful mechanism for controlling a relationship.

I suggest to beginners that following the opening "*How do you feel?*" it is challenging to try to facilitate the patient's narrative without using another interrogative. This is so different from what they have learned in medical school—the green or blue book with the hundreds of topics about which the patient must be questioned—and there is a strong pull to retain that interview structure. It is familiar, safe, and very much the doctor's role. "How," "when," "where," and "why" come naturally to mind as one listens to the patient.

Perhaps the single strongest facilitator of narrative flow is the interviewer's genuine interest in the patient's story. This is communicated in many ways, most of which are not verbal. Most of us have had the experience of telling a story in a social situation and felt the braking effect of the listeners' looking away or other signals of disinterest. For the beginning resident, the task is to relearn a natural stance, to put aside the expressionless pseudo-objectivity of the medical interview and to allow one's self to resonate to the patient's story.

Another powerful stimulus to storytelling is having a listener respond to the affective component of the narrative with simple empathic statements. Such responses lead the storyteller to feel understood, and understanding encourages and deepens narrative flow. This issue, discussed in earlier chapters, is important because many students of psychotherapy believe that the patient's level of experiencing during sessions is a predictor of positive outcome. Experiencing means the level of emotional involvement in one's own narrative. Although doubtlessly there are exceptions—such as the affectively overly expressive patient who needs greater distance from his or her experience—for many patients, the greater the level of affective involvement, the greater the likelihood of learning. It is, then, crucial both to encourage students to facilitate narrative flow and to facilitate deeper levels of experiencing.

General encouragers are brief phrases that encourage the patient to continue. They are spoken with clear interest, but without an authoritarian tone because it is important that the patient not feel ordered to proceed. General encouragers are not usually responsive to a particular theme, but are meant to encourage the overall flow of the narrative. In this way, the interviewer has the opportunity to note the patient's associative linkages.

When the patient stops talking, the interviewer has several options. One can sit quietly and wait for the patient to resume or, after several moments, use a general encourager such as "Go on" or "Please tell me more." At times, however, presenting the patient with a brief summary of the patient's narrative acts as a facilitator of further movement. Such summaries are brief and, in effect, let the patient know that one is listening carefully.

I introduce the students to the use of projective statements as discussed by Havens (1986). He points out that the issue is to "set the other in motion as naturally and richly as possible" (p. 100). For some patients, questions lead to the feeling of being interrogated; for others, empathic statements may suggest a closeness that is experienced as dangerous. Projective statements are declarative statements of fact or possibility that are meant to evoke. The patient can accept, amend, or reject them. By their use, the therapist demonstrates his or her willingness to be wrong; as Havens notes, "he is delighted to be corrected and have the record set straight" (p. 108). Projective statements are usually made in an offhanded way; they are meant to evoke, not challenge.

"There was no way I could do it. . . ."
  "No way at all."
"I wanted to help, I really did. . . ."
  "To want so badly. . . ."
"There was no way."
  [silence]

Here the therapist's empathic responses do not facilitate the patient's narrative. After 20 seconds, the therapist makes another empathic statement, this time focusing on the patient's resistance.

"It is so hard to talk about."
  "Yeah."
[silence]

The therapist notes that the empathic route does not work and after a brief pause, offers a summary.

"Let's see. You were only 13 and after your sister's death your mother just gave up. You wanted to help, wanted badly—but there was no way."

"Why did she do it? Why did she? She knew I had no one else [with increasing feeling]*."*

Noting that the brief summary did what the empathic statements did not, the therapist maintains the distance and offers a series of projective statements.

"She should have known."

"Anyone would know, I was only 13 . . . what did she expect . . . for me to be able to care for myself?"

"She left you all alone."

"And ever since it's been that way. Even when I lived with Polly it wasn't right, there was a part of me . . . somehow I'm always alone. I keep looking somehow for a person . . . or something to make me feel less alone."

This brief exchange illustrates both the use of several methods of facilitating narrative flow and the concomitant search for the metaphorical distance that will enhance the process.

The exercises used to help students learn to facilitate narrative flow start on the first day of the seminar. The initial exercise is the demonstration by my role-playing of the differences between a collaborative exploration and a directive inquiry. The resident's role-playings are captured on videotape and allow the group to review each participant's efforts. Invariably, residents are surprised at how entrenched the directive inquiry has come to be, how much they rely on the use of questions, and how difficult it is to learn an interview process of a very different structure. Repetition and practice are essential.

From the beginning, attention is directed to the recognition of recurring patterns in the patient's narrative. This issue is a central piece of the process of a psychodynamic formulation, itself a part of a broader synthesis, the clinical formulation. The process and format for constructing a clinical formulation are presented in the next chapter.

I am not certain what has changed about my use of confrontation, clarification, and interpretation since my training and early years of practice. It may be that the change is mostly in the professional set about the nature of psychotherapy and its crucial processes. The change could be in the nature of the material brought by patients to their therapy. And there is a real possibility that the change is my own, part of the process of professional maturation and personal aging.

When I was in training, the use of the word "therapy" meant only one thing; psychoanalytic psychotherapy. Nowadays, there are hundreds of widely differing forms of psychotherapy, some of which are considered teachable by manual. Further, the psychoanalytic psychotherapy of my youth seemed mostly a unitary thing, rather than a host of different forms. The metaphor that best described the process of that psychotherapy was surgical—a kind of excision by interpretation (Havens, 1986). Psychopathology was a foreign body, not of the same order as the person containing the foreign object. It could be dissected—ruthlessly if necessary—and clarification, confrontation, and interpretation were the probe, hemostat, and scalpel. Thus, the metaphor contains the aggression thought to be required to achieve the cure. It is no wonder that we confronted, clarified, and interpreted boldly without doubt as to the absolute reality of the issues being dissected.

The intellectual climate of mainstream psychotherapeutic psychiatry seems vastly different today. If the surgical metaphor continues to exist at all, it is probably in the hands of strategic therapists, full of manipulative cunning and paradoxical interventions. Psychoanalytic psychotherapies have different metaphors, and they range from containers to mirrors. It is not that confrontation, clarification, and interpretation are not used; rather, it is the spirit in which they are used that has changed. Most often, the central issue is how to use them within the context of a collaborative relationship—in effect, how to encourage the patient to do his or her own confronting, clarifying, and interpreting.

Actually, it is more complicated. The issue is how to use the power of such interventions with different patients having very different levels of psychopathology. Today's therapist must consider with whom power is better shared and with whom its use is at least somewhat surgical. My perspective on this complex issue is that it is wiser to search for a collaborative relationship and use one's power surgically only in those clinical situations in which a collaborative alliance is impossible to attain.

Has the nature of psychopathology changed in the 40 years of my experience or are we looking at the same processes from a vastly different perspective—thus deluding ourselves that real change has occurred? I vacillate on this one. Much of the time it seems to me that the change is real—that there is nothing now like the hysterical paralyses and other oedipal presentations of my youth. Nowadays the preoedipal reigns, and the separation-individuation metaphor is omnipresent.

At other times, I am convinced that nothing has changed but the conceptual lens through which we filter our observations. All the hysterics of my youth were really preoedipal and, if the separation-individuation lens had been available, it would have influenced that view of psychopathology.

I suspect there is some validity to each perspective, and from one view it doesn't make much difference which is most accurate. That is because we see, or think we see, more primitive psychopathology, the kind that does not readily respond to cognitive interventions alone and requires more holding or containment. Thus, time has blunted the power of our surgical instruments. Despite DSM-III-R and other attempts to establish clear syndromic boundaries, much of contemporary psychopathology presents as longitudinal patterns of difficulties relating to self and the world. In such a situation, the use of confrontation, clarification, and interpretation is most often a softer art.

Women are said to become more authoritative as they age, and men to become more collaborative. Once again, I thought I saw this in my father and his generation, but almost certainly it is more complex, for all things look different at 40 than at 20. Nevertheless, I suspect that there is some truth to the axiom, and my generation is more mellow in our 60s than we were in our 40s. More mellow has something to do with the use of power and, in particular, ease in sharing it with others. Although we are still capable of exquisite obstinacy, for the most part it seems more effective to rely reciprocally on others.

Some of this no doubt reflects the impact of 25 years of family research and, in particular, the studies of well-functioning marriages and families (Lewis, 1988, 1988a; Lewis, Owen & Cox, 1988). The spouses' effectiveness in such families seems clearly related to their joint capacity for collaboration, their readiness to facilitate each other's leadership.

Whatever the complex play of changes in psychotherapy, psychopathology, and myself may be, my thinking about and teaching the process of confrontation, clarification and interpretation is different today than it was 20 years ago. The theme now is clearly one of encouraging students to think about how to use these interventions with different patients, but the emphasis is on their use in a collaborative manner.

Clarification, confrontation, and interpretation are introduced early in the seminar because not infrequently the exercises (role-playing, audiotaped patient stimuli statements, and videotaped patient stimuli statements) used to teach simple empathic responses also contain indications for one of these more cognitive interventions. Thus, a video segment of a middle-aged woman saying in a pleasant manner that she wished she were dead, or an audio segment of a man's voice telling the doctor in a matter-of-fact fashion to go to hell raise the issues of incongruent messages and the use of confrontation.

The most common incongruent messages are like the two examples given above: the words and affect do not match. When intense or alarming messages are spoken blandly, attention is drawn to the denial of affect. A confrontation (called "everyday" confrontation to distinguish from "heroic" confrontation) aims to undo denial, or at least to test the feasibility of doing so. Looked at from a different perspective, it may be, as Havens (1986) suggests, that denial of affect always depends upon the collusion of another person. From this perspective, the interviewer is refusing to collude.

The suggestion that the clinician confront denial even in the initial moments of the first interview is often viewed with alarm by beginning students. Their concern is that the patient may interpret the confrontation as criticism or an indication of the clinician's oppositionalism. I suggest to them that there are several reasons why confrontation can be used effectively right from the start. First, it helps to establish the clinician's emphasis on openness and honesty. Secondly, not to do so runs the risk of suggesting one is not interested in the patient's affective experience, or is even put off by it. Finally, the manner in which the confrontation is made is important. I suggest that initial confrontation be made in a way that makes it relatively easy for the patient to disregard the confrontation. Tentativeness is often useful in this regard.

"*It sounds dreadful, but it's almost as if you're really calm about it.*"

"*I hear you saying that you're okay, but I thought I saw your eyes glisten with tears.*"

"*You say you're not angry, but you seem to be gritting your teeth.*"

The use of "*almost as if*" "*thought*" and "*seems as if*" presents the confrontation in a manner that allows the denial to remain intact if the need for it is sufficient. The idea of respecting the patient's defenses is an old one, and the critical issue is how to know when to respect them and when not to. Being tentative can be understood as an initial probe or trial balloon.

The most fundamental act of clarification is the request. "*Help me understand that better*" or "*What was that like for you?*" are examples in which the patient is asked to make the clarification. A greater level of activity on the part of the clinician is reflected in "*It could have been like that time when you were a little boy . . .* " Here the clinician is offering the patient a piece of detail that might clarify a current situation. "*Could have been like . . .* " suggests the hypothetical and is used to make it easier for the patient to decline the offering. "*There is something familiar here, but I can't quite make it out*" is an even softer attempt to encourage the patient's effort to clarify.

The intent of these statements is to encourage the patient to do some of the work of clarification or, if possible, most of the work. The more the patient experiences the clarification as his or her own work, the greater the potential impact of the revelation, I believe. If the clinician consistently does most or all of the work, the relationship is assuming a parent-child configuration. Although some patients may be best served by that type of relationship, most are not.

Interpretation has been the heart of psychoanalytic psychotherapy since its very beginning. It has been the big gun. Telling another person what he or she says or thinks or means is powerful stuff. It never was quite that way, of course, for an interpretation was only that invasive in the extreme. I recall an interpretation made by a supervisor who said (in a group setting!), "*Lewis, you will never deal with that material effectively until you return to the Catholicism of your youth.*" He blew me away, in part because I thought his interpretative comment was grossly inappropriate for the context, but also because it had an uncanny element of validity . . . not about Catholicism but about my Roman Catholic mother and our relationship. Its impact was powerful.

Interpretation, like confrontation and clarification, has softened over the past 40 years. If it ever really was Moses from the mountain top, it is no longer . . . at least, not often. For the most part, it is a collaborative effort in which the therapist hopes to facilitate the patient's interpretive ability. Further, there is increasing realization that interpretations serve many purposes in addition to insight. Kramer (1989) describes some of these purposes, including interpretation as a source of support, a message that the patient is important to the therapist and that his or her experience is comprehensible. Depending on how the interpretation is made, it may communicate the therapist's firmness, self-assurance, power, or steadfastness in the face of initial defenses. It can be used to stimulate the patient's interest, to set an example about how to think about a problem, or, conversely, to produce confusion.

As Kramer points out, the impact of an interpretation is often shaped by how it is done. Under most circumstances I want the patient to uncover the connection, to feel that it was his or her work that counted. For all but a few patients, I want them to feel a step ahead of me in the development of understanding. The impetus is thus for a collaborative effort in which the patient increasingly takes the lead. "*How do you understand that?*" is a prototypical beginning to an interpretive interchange. "*Does it seem anything like those years in high school?*" points a direction; it offers more, but stops short of making the explicit connection.

Not infrequently a patient will ask if I knew in advance about a connection he or she discovered.

"*I've been thinking since our last session about my weight.*"
 "*Trying to figure it out.*"

"Yeah. You know, since her death . . . it's like I've been feeling deprived, and I'm darned if I'm going to deprive myself . . . so I gorge."

"Eating is connected to feeling deprived . . . that's really interesting!"

"It really is something that just came to me . . . but you've known it all along, haven't you?"

"This one is yours."

"Maybe . . . it builds upon all we've been talking about."

I have emphasized the use of these common cognitive interventions in the context of developing a collaborative relationship in which the therapist's socially sanctioned power is deliberately shared with the patient. Those circumstances in which the therapist wishes to retain or augment his or her power are not rare, but clearly are the exceptions. It is important to help beginning students to understand that how a therapist uses confrontation, clarification, and interpretation is determined by many factors. At the heart of the decision-making, however, is an understanding of the patient's psychopathology. Treatment is based on this understanding and, thus, it is important to introduce students to an approach to synthesizing the interview data into a clinical formulation.

CHAPTER 7

The Clinical Formulation

My teaching students the art of the clinical formulation begins with an introduction to the dialectic concerning psychopathology. The mainstream view is empirical and holds that psychopathology—at least in its more severe forms—is a biopsychosocial reality, real, palpable almost in its concreteness. Severe psychopathological states have genetic diatheses, clear onsets, more-or-less typical untreated clinical courses, and predictable endings. They can be studied using state-of-the-art scientific methods and ultimately will yield to an accurate appraisal of the complex interplay of many causative variables.

Another perspective, the constructivist, or narrative, orientation holds that what is central is the nature of human experience. We experience our lives mostly in linear terms, and the narratives we construct usually follow whatever forms are most approved by the culture. From this perspective, psychopathology involves the imposition by presumed experts of more-or-less artificial criteria in order to bring to manageable form the otherwise infinite number of categories of psychopathology. Psychopathology, even in its more severe types, can be categorized only because of the power of the consensus of presumed authorities.

This dialectic is not easy for most beginning residents. Just as they finish their internship experiences, when an initial sense of biomedical competence is at hand, they enter the world of psychopathology and feel, once again, an encompassing shroud of incompetence. An empirical approach is much more like thinking in general medicine—its familiarity reassures. It seems clearly to be where and how to start beginners thinking about psychopathology.

I share with residents something of the dialectic within me, a dialectic that, despite involvement in empirical research since medical school days, enchants me. I suggest that they, too, may someday find the enchantment of conflicting views of reality but, for now, the emphasis is to be on what is demonstrable and how best to formulate it.

The philosophical debate about the most appropriate perspective from which to view psychopathology raises the question of model-building as a fundamental human characteristic. This view states that it is impossible to be without a causal model when viewing or participating in an event of some consequence: the question is whether the model is to be explicit or implicit. Whether explicit or implicit, such internal models can have profound impact on the lives of individuals.

Recently, Seligman (1989), in reviewing research regarding explanatory style, asks what a person brings to or projects on a situation that either makes him or her vulnerable or invulnerable to helplessness. If one responds to negative events with an internal, stable, and global explanation ("*It's me. It's going to last forever. It's going to undermine everything I do*"), one is more vulnerable to depression, ill health, and lesser achievement than if one's explanatory style is different (*It's not me. It's not going to last forever. It's only one part of my life.*").

Explicit models are less dangerous than implicit ones because unacknowledged causal inferences are more prone to reflect the individual's underlying biases and prejudices, which then go unexamined. Making a model explicit often brings into clear relief the operation of one's biases and prejudices.

At the same time, it must be acknowledged that all causal models are embedded in and reflect the sociocultural and historical contexts. For example, definitions of normality are molded by cultural value systems. Currently, autonomy, the ability of the individual to function effectively on his or her own, is emphasized as the hallmark of normality or maturity (without equal emphasis on its complement, the capacity for intimacy). This emphasis surely reflects, in part, this culture's valuing of individual achievement and successful competition.

Another perplexing issue is how complex a model of psychopathology should be. Is there an optimal level of complexity for all clinical situations, or does the clinician need several models of differing complexities so as, for example, to deal with acute and chronic clinical circumstances? Currently, much attention is paid to complex models of psychopathology, models that incorporate biological, psychological, and social variables. Haley (1980) suggests, however, the usefulness of distinguishing between models of family dysfunction for research and models upon which to base clinical interventions. Research models seek to explain all the variables that, in concert, produce the clinical situation, whereas clinical models aim to produce change and are often relatively simple.

Currently, advances in neuroscience have fostered biological intervention models that direct treatment to the modification of presumed abnormalities in various neurotransmitter systems. These relatively simple models pay scant attention to psychological or social variables. Thus, severe psychiatric syndromes are labeled "brain disease."

Engel's (1980) biopsychosocial model is currently the most attended complex model of psychopathology. Its beauty is found in its breadth, ranging from the biosphere to subatomic particles, but it does not provide clinicians with clear guidelines regarding the appropriate level (or levels) at which to intervene. This may lead clinicians to intervene at the level with which they are most familiar. As I look back at my clinical career, the evidence for what Havens (1973) calls the "mad surgeon" syndrome* is obvious. I was trained to treat patients with psychoanalytic psychotherapy and to intervene at the level of the person. With rare exception, I treated each patient with this modality. Only later, as I learned about marital and family systems, did it become obvious that there were choices to be made, choices about the level of intervention and the modality to be used. During earlier years, however, I was like a "mad surgeon."

Beginning students must learn how to do a clinical formulation and learn how to think about clinical data—which data to include, how to organize the data, and how to integrate the data in a way that provides a model of the factors that explain the patient's dilemma. Beginning students must be encouraged to think broadly

*A "mad surgeon" is a surgeon who removes each patient's gall bladder regardless of what is wrong because that is the only surgical procedure he or she knows how to do.

about which variables are relevant. The phase "clinical formulation" is used to promote consideration of biological, psychological, and social system variables. The psychodynamic formulation is but one part of the clinical formulation, necessary but not sufficient. A DSM-III-R diagnosis is also necessary, but it is only one part of the clinical formulation. The outline used in the seminar is shown in Table 7.1.

TABLE 7.1.
Outline for a Clinical Formulation

1. *Descriptive Statement.* A brief statement describing the development of the patient's syndrome emphasizing patterns as well as episodes.
2. *Developmental Data.* A statement including the status of the patient's family of origin at the time of the patient's birth, the patient's temperament, and childhood, adolescent, and adult developmental markers.
3. *Biological Data.* A statement attending family genetic information, the presence of physical disabilities, serious illnesses, and endocrine disturbances, information regarding alcohol and drug use, head injuries, and any neuropsychological data as well as the results of special tests, including neurological examinations, EEG, and other procedures. Data from the mental status examination suggesting cognitive impairment should be included in this section.
4. *Psychological Data.* A psychodynamic formulation that permits considerable variation depending on the clinical context, amount of time available for the assessment, and the level of the clinician's experience. A central theme, conflict, or organizing metaphor is identified from the patient's narrative. Opposing motives and wishes are elaborated using one or parts of three contemporary models: ego psychology, self-psychology, and object relationships. Particular attention is paid to the maturity of the patient's defenses.
5. *Social System Data.* A social system formulation includes the patient's participation in ethnic, cultural, social class, religious, and community systems. Particular emphasis is placed on the immediate interpersonal context; the patient's current family system and his or her participation in a central dyadic relationship.
6. *Transference and Countertransference.* A statement is made regarding either noted or anticipated transference and countertransference themes.
7. *Personality Strengths.* A brief statement about the patient's strengths, including intelligence, persistence, capacity for self-observation, openness to new experiences, and ability to postpone gratification is necessary.
8. *Value Orientation.* A brief statement is made about the patient's core values, with emphasis on particular cultural, religious, and existential value systems.
9. *DSM-III-R Diagnosis.* The descriptive data are formulated in terms of the DSM-III-R categories.
10. *Prognosis.* A brief statement is made about the clinician's judgement regarding the patient's prognosis with and without appropriate treatment. Reality factors that interfere with or preclude appropriate treatment are to be noted.

1. *Descriptive Statement.* The brief descriptive statement is important because it is sometimes possible to note recognizable patterns of psychopathology over time.

> This 18-year-old adolescent has always been shy and withdrawn. An excellent student and full scholarship recipient, he became even more withdrawn as time to leave for college approached. One week before his anticipated departure he began to experience auditory hallucinations informing him that he was queer and delusions that he was the object of a drug cartel's secret plot to overthrow a South American country. He became progressively more agitated and was admitted to the psychiatric unit of a general hospital.

Although this statement contains all sorts of suggestions about the patient's clinical condition, it raises the question of a major psychotic process evolving in a perhaps schizoid adolescent at the time of separation from the family.

2. *Developmental Data.* The developmental data start with whatever information is available about the family of origin at the time of the mother's pregnancy and the patient's birth. An array of factors often impinge upon the family at such a time: whether the pregnancy was planned; whether the pregnancy occurred at a propitious time for the family; the circumstances of the mother's health prior to and during the pregnancy, as well as during the postnatal period; the nature of the patient's birth, with particular attention to complications; the patient's immediate physical health; expectations and feelings about the patient's gender; the family's identification of the patient with a particular relative, living or deceased; and other special circumstances that may color the emotional climate of the family during the period preceding and following the patient's birth.

It is important also to search for information about the patient's temperament, and the work of Thomas and Chess (1977) provides a useful descriptive system. It is sometimes possible to obtain clues regarding the "fit" between the child's temperament and each parent's personality.

These factors provide the crucial context for the more classic developmental markers of childhood and adolescence. The rapidly growing literature on the developmental stages of adulthood offers the clinician useful constructs for understanding post-adolescent development. The developmental data for the 18-year-old adolescent boy include the following:

> The patient is the only child of a middle-aged couple who had tried unsuccessfully for 12 years to have a child. There was great expectation that the patient would be male because otherwise the father would be the last male to bear his family name. The parents were described as isolated socially, somewhat enmeshed in both families of origin, and relatively affluent. The father was clearly dominant, but the mother seemed satisfied with her less powerful position. The pregnancy and birth were uneventful. From the start the patient was sickly, and the mother's delight in him soon forged a bond that excluded the father. The patient was quiet and shy "from the start" and had few peer interactions.
>
> The classic developmental markers of childhood were reported to be normal; early school experiences suggested a moderate level of school avoidance. Puberty was delayed, the patient had no strong adolescent friendships and had not demonstrated any interest in girls. His management of sexual urges during adolescence is unknown.

3. *Biological Data.* The search for biological factors needs to be wide-ranging, and particular emphasis is placed on potentially reversible processes such as endocrine disturbances, subdural hematomas, misuse of prescription medications, and use of illicit drugs. The increased use of neuropsychological tests, EEGs, and more recently developed special procedures such as scans and beams has brought greater awareness of the role of often subtle cognitive deficits. Frequently, there is no history of head injury or other occurrence that might have been causative, and the clinician may be perplexed as to the origin of these deficits. Although more gross neuropsychological deficits may clearly limit what is possible in treatment, the clinical relevance of the more subtle deficits may be impossible to ascertain.

Among many clinicians sensitivity to the possible role of subtle temporal lobe syndromes has increased, and any suggestion of rhythmicity in symptoms leads to the consideration of a trial of anticonvulsants. We live in an era of aggressive searching for possible biologic etiologies, and there is no question that such searches have resulted in spectacular improvements for some patients.

> The adolescent patient is not known to have used drugs of any sort. There is no history of head injury. Thyroid studies were reported as normal. There is, however, a clear multigenerational family history of the presence of extreme shyness and introversion. A great grandparent on each side of the family is reported to have died in a state hospital after many years of hospitalization. In the absence of evidence suggesting organicity from the mental status examination and a psychological test battery, neuropsychological tests and other specific procedures were deferred.

4. *Psychological Data.* For the beginning resident the psychodynamic formulation presents a number of problems. First, a level of inference is required to move from descriptive observations to presumed underlying wishes, conflicts, fears, and defenses. Second, there is an imposing array of systems of formulating: ego psychology, object relations, and self-psychology. Third, residents often hear different formulations of the same patient from different senior clinicians.

Because of these problems, the seminar starts with a task requiring little inference. The residents are asked to identify interpersonal episodes from the patient's narrative and then to identify what Luborsky (1977) calls the Core Conflictual Relationship Theme (CCRT). The CCRT involves the patient's wish, need, or intention, and its consequences both from the other person and from the self. This system of identifying a recurrent conflictual theme has been extensively studied and is similar to the approach of the Vanderbilt Psychotherapy Research Group's Cyclical Maladaptive Pattern (CMP). Both of these approaches are research-based and easily understood.

In the beginning, it is wise to use verbatim transcripts of interviews and ask the students to score the transcripts for interpersonal episodes and CCRTs. They quickly master these tasks and then move to exercises involving videotaped interviews. Later, they are asked to do the same thing with the exploratory interviews they are accomplishing with patients.

Next, the residents are introduced to the three basic types of psychodynamic formulations through the work of Perry, Cooper, and Michels (1987). It becomes clear to them that the ego psychology, self-psychology, and object relations models are complementary ways of conceptualizing psychopathology. The ego psychology model is particularly suitable for neurotic level psychopathology. It is the oldest form of psychodynamic formulation, emphasizing the role of unconscious wishes and fears, the conflict between them and superego or reality, the central role of the resulting anxiety, the use of characteristic defenses, and the resulting symptoms, inhibitions, and character traits. The oedipal conflict is central; preoedipal conflicts and fixations and interpersonal variables are not particularly emphasized. Anxiety and conflict are the important constructs. This model of psychodynamics is entirely intrapsychic, and it deals primarily with the derivatives of sexual and aggressive drives.

The self-psychology model emphasizes the development of the self-system. Its focus is on the child's expression of strivings and ambitions and, most of all, on the empathic responses of parents and others. The internalization of empathic others and the consequent development of a firm sense of self with its associated capacity for joy, creativity, and empathic relationship are central. The empathic failures of others, with associated distortions in the self-system and subsequent defensive orientations, lead to the development of psychiatric syndromes. This model finds particular utility in conceptualizing a wide range of narcissistic disturbances. In its emphasis on the empathic responses of significant others, self-psychology moves the focus toward the interpersonal world of the patients.

The object relations model emphasizes the internal representations of self and others interacting. Thus, it is another step in the direction of a more interpersonal orientation. Internal representations range from primitive to realistic; each is associated with a primary affect and wishes and fantasies. "Good" and "bad" self and object representations, originally seen as split into separate organizations, develop into integrated representations that maintain optimal self-esteem. The projection of unconscious "bad" self and object representations is a central construct for severe psychopathology and, thus, this form of psychodynamic formulation is most useful for patients with psychotic or borderline syndromes.

The residents are also introduced to a comprehensive model of a psychodynamic formulation as presented by Friedman and Lister (1987). This model includes characterologic, dynamic, and genetic constructs. Finally, they familiarize themselves with another perspective on early development, Stern's (1985) description of the development of a sense of self during the first 30 months of life.

Teaching beginning residents various approaches to the psychodynamic formulation is like offering a smorgasbord. I wish to introduce a variety of approaches, encouraging the residents to select the one (or a combination of elements from several) that best suits each patient, a way of understanding that translates into clear treatment dimensions. The psychodynamic formulation is not a search for an exalted truth; rather, its purpose is to direct treatment.

> Exploratory interviews with the 18-year-old patient produced a halting, often fragmented narrative. Interpersonal episodes he recalled were either very brief or lengthy and tortuous. It was possible to identify a CCRT that emphasized either being in a commanding position in relationships with others or being an impotent pawn manipulated by powerful others.
>
> The patient's psychopathology can be understood as being fixated at a basic and primitive level of incomplete individuation from his mother. His self-system is fragile and incompletely integrated. His underlying anxiety is severe and may reflect a fear of complete fragmentation and ensuing nothingness. Characteristic defenses centered about schizoid withdrawal, denial of most affects, and rigid, obsessional approaches to most situations.
>
> The patient's psychotic regression can be understood as a response to his coming separation from his mother. The process of his psychosis is characterized by alternating periods of delusional grandiosity and periods of helpless victimization.

5. Social Systems Data. The interpersonal perspective is summarized by noting the characteristics of the social systems in which the patient participates, and the role or roles that he or she plays. Some sense of order may be attained by starting with the larger

social systems and moving then to the smaller and more intimate systems. Ethnicity, culture, and class are the "big" systems, and their appraisal begins the process of orienting the beginning clinician to the patient's social context. Community, neighborhood, and peer group systems are of intermediate size and often offer important information about factors that appear to influence the behavior of individuals more directly. Family and marital systems are the day-to-day stuff of life, and their molding and shaping impact upon the individual can be enormous.

At each system level, the clinician needs to know the prevalent values and something of the system's structure: how organized or chaotic; how rigid or flexible; how isolated or permeable; how power, distance, closeness, and intimacy are organized; how the past, the present, and the future are emphasized; how much emphasis is placed on the individual and how much on the group; how success and failure are measured. These and other critical dimensions of social life need consideration.

Although assessment of each of these systems is important, systematic appraisal of the patient's marital and family systems usually dominates the clinical formulation because powerful treatment modalities are available at the level of the smaller systems and the clinician must often face the choice of individual, marital, or family level of intervention. Intervention in community or neighborhood, let alone in the truly big systems, is beyond the ability (and power) of the clinician.

Thus, I emphasize the need to appraise carefully the patient's marital and family systems as a crucial part of the clinical formulation. Because these smaller systems are an important part of my clinical and research activities and because they are not often addressed in any detail in the usual clinical formulation, I will describe their appraisal more completely in the next chapter. To return to our patient:

> Both the maternal and paternal families are of Western European ancestry, and both families have lived in this country for many generations. The families have been, for the most part, educated and relatively affluent. There is a history of small family size, extensive contacts with the extended families, and relatively little social involvement with the outside world. Shy-

ness is an individual characteristic present in at least one member in each generation. The family structure is relatively impermeable, enmeshed, and isolated.

The parents' marriage is dominant-submissive and complementary, as the less powerful wife seems satisfied with her role. The patient's father is a very controlled, obsessional man, a successful investor with no outside interests. The patient's mother is a somewhat shy woman with carefully modulated affect. Her relationship with the patient is an intense coalition in which she continues to treat him like a young child. This cross-generational coalition seems acceptable to the father as it allows him more time alone to study his investments. The patient appears to accept his mother's age-inappropriate treatment without complaint. If she fails to lay out his clothes each morning, he asks her to make the decisions.

Although this is a quiet, enmeshed, and clearly dysfunctional family system containing a pathological triangle, there is no suggestion of primitive, bizarre cognition. It is clear, however, that the adolescent son receives nothing from either parent that facilitates the separation-individuation process.

6. *Transference and Countertransference*. After the initial interview (or several) the clinician can formulate preliminary ideas about the probable nature of the patient's transference and the clinician's countertransference. These constructs often confuse beginning students, in part because they are sometimes defined in a relatively narrow and specific way and at other times in a more global fashion.

Helping students to learn more about their usual interpersonal impact is the teacher's initial task and is aimed at helping them to distinguish the patient's feelings and thoughts that the interviewer's behavior invites from those that seem more clearly projections from the patient's past relatively uninvited by the interviewer. The early videotaped role-playing and interviews with actor "patients" proved the most useful format. Each student's colleagues are encouraged to be helpful, and gradually their comments become more specific. Body size, posture, facial expressions, and other aspects of the student's behavior are noted.

The introduction of the distance-regulation metaphor and the subtle negotiations that can be understood as a search for an optimal working distance provide a rich vein of information. Of particular help are those occasions when it is clear to all participants that the student interviewer retreated from the patient's experience. Throughout these exercises, it is important for the instructor to model making observations in a nonjudgmental way. When such an attitude is established as a group characteristic, the students' resistances to these observations diminish or disappear.

Teaching students to recognize early transference projections from the patient is made easier by the use of the Core Conflictual Relationship Theme (CCRT) and the Triangle of Insight (Butler & Binder, 1987). The identification of a specific relationship theme in both a current central relationship and a childhood relationship serves as a harbinger of the wishes and conflicts the patient may bring to the treatment relationship. On some occasions, there is evidence in a single interview of the same theme in the past, the present, and the transference relationships.

The students are introduced to countertransference in a variety of ways. Viewing their videotaped interview helps them not only to understand their usual interpersonal impact, but also to learn what affects, themes, and conflicts they wish to avoid. The Forced Fantasy Exercise also highlights specific issues that are processed idiosyncratically.

Each exercise contains the message that it is necessary to know one's self at a deeper level, a depth unnecessary in other areas of medicine.

Two other processes help in the search for self-knowledge. The first is construction of a careful description of their families of origin, in particular their specific roles in the families. They are not asked to make public their family of origin reconstructions, but I emphasize that many roots of their countertransference propensities can be clarified by a closer understanding of their roles in their early families. Second, by using fantasy, beginning students discover that either spontaneous or induced fantasies about patients can provide a remarkable window into their selves. During the course of the seminar, their reluctance to report fantasies about patients gradually wanes. The resistance never completely disappears, but it becomes less compelling. For some students the use of their fantasies beginning in these early days becomes an important aspect of their work with patients.

> The therapist who began to see the 18-year-old adolescent during the early days of his hospitalization reported a central concern with whether a therapeutic alliance could be established. She also reported some concern with whether the unit team clearly understood how ill the patient was, and whether they were sensitive enough to the malignant anxiety she believed was just beneath the patient's psychotic symptoms. She reported also that the charge nurse had raised the question at a team meeting of whether the resident wasn't increasing the dosage of the patient's psychotropic medications too rapidly.

Each of these issues is relevant in and of itself. All must be dealt with, however, at two levels. The first level is the observable ebb and flow of early inpatient treatment with a very disturbed youngster. The second level is that of countertransference, the therapist's need to care for the patient in a special way—a way that repeats not only certain aspects of the patient's experience with his mother, but also certain aspects of the resident's experience in her family of origin.

7. *Personality Strengths*. Because the focus of most of their thinking is the patient's pathology, beginning students need to be encouraged to review each patient's strengths. Strengths include intelligence, ability to persist in accomplishing tasks, capacity for self-observation, ability to postpone gratification, and openness to new experiences. Such traits are clearly attributes of healthier or more mature ego functioning and are often lacking in many more disturbed patients. Nevertheless, it is important to consider strengths in each patient's formulation.

Strengths also can exist outside the patient's personality, such as membership in marital and family systems that support individual psychological growth rather than systems that require an identified patient. In a similar way, the presence of support groups can be a strength for the patient.

> The resident's formulation of her 18-year-old patients's strengths noted only the evidence of his high level of intelligence. Evaluation of other personality strengths was clouded

by his psychotic functioning. The patient's family system did not facilitate his continued development as a clearly differentiated individual, and his schiziod nature was reflected in the almost total absence of relationships outside the immediate and extended families.

8. *Value Orientation.* A careful consideration of the patient's core values is an important aspect of the clinical formulation. Students need to know what beliefs patients hold as central to their lives, and the ways patients' values differ from their own. Often such differences are cultural or religious. When clinician and patient come from different ethnic or cultural traditions, it is imperative that the clinician carefully monitor the ways in which they may differ about what is important in life. A classification scheme such as that of Spiegel and Kluckhohn (1971) is a useful format to help organize residents' thinking about value orientation. The varying emphases of different groups regarding nature, activity, time, and relational orientations may illuminate differences in the values of patient and clinician.

Perhaps religious values receive the least attention by many clinicians. It is rare to hear mention of a patient's religious beliefs and practices during a teaching case conference. This is particularly unfortunate since evidence suggests that many psychotherapists do not share the belief of most patients in a supreme being, salvation, and life after death. The psychotherapist's awareness of these differences and respect for the patient's values can be crucial for successful treatment.

To this end, each resident completes a Rokeach Value Survey during the early days of the seminar. This is the starting point of a focus on values that continues throughout the seminar. In addition to cultural, religious, and existential values, the seminar emphasizes an appreciation of institutional values. For example, what values does an institution (department of psychiatry, psychiatric hospital, etc.) espouse? Is the institution's behavior congruent with its stated values? Are those values reflected in the institution's educational activities? Such issues bring to real immediacy the often abstract concept of values.

Little is known about the 18-year-old adolescent patient's value systems. His family appears to reflect conservative middle class values. The enmeshed quality of both the immediate and extended families may be interpreted as reflecting an underlying belief that the world is dangerous, but this remains highly inferential.

9. *DSM-III-R Diagnosis*. The development of effective psychotropic medication for a growing number of syndromes has increased the importance of accurate diagnosis. This seems particularly so for Axis I disorders that are episodic in the sense of having a clear pattern of onset, easily distinguishable behavioral change, and a discernible ending. An episode of major depression is the prototype. Axis II disorders are far less responsive to medication, and accurate diagnosis is more difficult. The residents receive a separate course in psychiatric diagnosis and do not need to be taught the essentials in this seminar.

Although the 18-year-old adolescent patient's psychosis does not meet the DSM-III-R requirement for schizophrenia of six months of continuous signs of the disturbance, the treating resident believed that the patient's basic diagnosis was schizophrenia. She also believed that the patient had an Axis II disorder, schizoid personality disorder.

10. *Prognosis*. It is useful to ask residents to include a prognosis in the clinical formulation. Doing so requires them to review factors, positive and negative, that have been empirically demonstrated to be associated with outcome. Thus, factors such as good premorbid functioning, sudden onset under stressful circumstances, and a supportive spouse and family are positive prognostic indicators. A severe syndrome of chronic proportions, with a gradual onset, without clearly stressful circumstances in a patient from a severely dysfunctional family are the comparable negative prognostic indicators.

This section of the clinical formulation encourages the clinician to review also the current treatment outcome studies for various psychiatric syndromes in order that he or she be able to make a judgment about the probable effects of appropriate treatment.

The resident treating the 18-year-old boy thought that there were mixed prognostic indicators. The severity of the patient's syndrome, the schizoid personality structure, and the dysfunctional family were understood as negative prognostic signs. The acuteness of his psychosis, its relationship to the stress of separation, and the prevalence of positive rather than negative symptoms were seen as good prognostic signs. She concluded that the prognosis for the psychotic disturbance was relatively good; but the prognosis for alleviation of the schizoid personality disturbance was poor.

The resident felt that appropriate treatment included hospitalization, individual supportive psychotherapy, antipsychotic medication, and family therapy. This combination of treatment modalities is understood to be essential for the treatment of the psychotic process.

The use of a broadly based clinical formulation encourages students to look at an array of biological, psychological, and social system variables in understanding their patients' disturbances. I believe that such an approach is best during early clinical training because it reduces the likelihood that important etiologic variables will be missed. There are, however, disadvantages to the clinical use of complex models of causality.

First, current knowledge does not allow integration of the diverse biological, psychological, and social system variables. There are few bridging concepts. The biopsychosocial model provides a format for studying psychopathology from a broad perspective, but tells us little about how to integrate the rich array of variables.

Another disadvantage is the idea that complexity in thinking about causality can, at some point, be used to avoid intervention. The search in the clinical arena is for information upon which to base treatment rather than as a goal in and of itself.

A third hindrance for using a complex model of causality is that treatment indications are seldom clear-cut. If everything is interconnected, everything needs treatment. What determines treatment priorities? What should the timing of treatment(s) be? Such questions are not easily answered when a complex model of causality is used.

Another potential disadvantage is that complex models of psychopathology do not easily incorporate certain linear constructs such as individual responsibility. This issue has been of particular concern in the attempt to understand and treat violence within a marriage or in a family. If a prime tenet of a systems perspective is the interconnectedness of all processes, how can such a view incorporate the ideas of abuser and victim? Dell (1989) explores this area in an influential essay, and suggests that an adequate scientific explanation requires a systems view. Human experience, however, is cast in linear terms, and victims and abusers are clearly recognizable. Complex systemic models of psychopathology are unable to integrate linear experience.

Finally, another obstacle to using complexity is that, having created a model, one has a tendency to fall in love with it. For some people, simple models are more seductive (the lust for linearity), while for others the complex models are more endearing (the craving for complexity). Regardless of which type appeals to the individual clinician, the "in love" state is dangerous because it may prevent adjusting the model to fit the patient. Not altering one's clinical formulation as treatment progresses is evidence of the clinician's failure to attend new data. Such oversight suggests a countertransference problem (Lewis, 1979a).

Despite these difficulties associated with complexity and a systems perspective, there seems to be no other way to capture in our conceptualizations the richness and diversity of our patients' experiences. By consistently making a clinical formulation that includes the 10 categories I have described, a student of psychotherapy can keep his or her mind open, produce data with which to mark his or her progress as a therapist, and offer the patient the most comprehensive treatment.

CHAPTER 8

Marital and Family Formulation

In *The Lives of a Cell*, Lewis Thomas (1974) writes about societies as organisms, emphasizing our collective discomfort in understanding the individual as a part of a larger system. Yet a system of relatively few elements is capable of activity unattainable by the individual organism. For example, a solitary ant, whose nervous system has few neurons, is patently incapable of a thought, but a group of ants encircling a dead moth demonstrates a kind of thinking, planning, calculating intelligence. Why are such examples of separate animals cooperating to form a system totally unpredictable on the basis of the sum of the individuals' skills so discomforting to us? Why, in Thomas's words, do we humans seldom feel our "conjoined intelligence"?

As one part of a comprehensive clinical formulation, clinicians need to know the status of the interpersonal systems the patient participates in and his or her role in these systems. There is ample evidence to support the idea that under some circumstances one's spouse or family can drive one to drink or into madness.

This interpersonal perspective on psychopathology has never achieved mainstream status in psychiatry despite its American origins. A number of reasons account for psychiatry's relative inattention to these often crucial systems. At the broadest level, the focus of Western consciousness is on the individual. As a medical discipline, psychiatry shares general medicine's focus on the individual. The patient is the individual; relationship systems are not seen as functional or dysfunctional and capable of facilitating either symptom development or health. Current emphasis in neuroscientific research in psychiatry is starkly on the individual. Despite impressive documentation of the role of social variables in disease in both animals and humans, neurotransmitter systems are rarely studied in the context of relationship variables.

Psychiatry's way of thinking is also influenced by the politics of seeking support for research or for clinical endeavors. The promise of finding relatively simple biological causes and cures for major psychiatric syndromes, "magic bullets," invites support. Models involving many complex causes, the validity of multiple perspectives, and probability paradigms are not nearly as dramatic. Advocacy groups often oppose any view that implicates family relationships as of consequence in the etiology and course of psychiatric syndromes. Remembering how families were blamed for a family member's syndrome during the family therapy bandwagon, such groups now insist on the restrictive metaphor of brain disease.

Third-party payers discriminate against marital and family therapies as an approach to the treatment of even major psychiatric syndromes. Clinicians have little economic motivation to acquire such skills in clinical practice or to emphasize them in training programs.

Thus, a broadly acceptable format for conceptualizing social system variables has not become an integral part of the standard formulation of individual psychopathology. Because I believe that a crucial decision for the clinician is what level of intervention to use, it becomes imperative to teach students how to appraise and intervene at many levels.

As a consequence of the shortcomings of interpersonal diagnostic standards and procedures, I rely heavily on the marital and family research of which I have been a part at the Timberlawn Psychiatric Research Foundation and on my clinical work with couples and families (Lewis, 1979, 1989; Lewis, Beavers, Gossett, & Phillips, 1976; Lewis & Looney, 1983). In addition, whenever it seems suitable, I include the published work of many others.

Before discussing the format and procedures for a marital and family assessment, I will outline briefly the theory of marital and family systems I introduce to beginning psychotherapy students.

The marriage or family as a system is greater than the sum of its members, and its characteristics cannot be fully predicted by the study of the family members. A family has two central functions: to deal successfully with its context (the external function) and to facilitate the development of family members such that they are able both to relate closely to others and to function autonomously (the internal function). These central functions are value judgments about what the family "should" do and, as such, must be considered as applicable only in contexts in which physical survival is assured.

Families may be judged as functional or dysfunctional in the degree to which they accomplish these goals successfully. *Functional* and *dysfunctional* are global judgments of family effectiveness, but it is clear that *highly competent* families have to wrestle with a variety of problems, and *severely dysfunctional* families may have strengths.

The research at the Timberlawn Psychiatric Research Foundation supports the concept of a Continuum of Family Competence, a way of distinguishing a small number of family types according to their variable capacities to accomplish the primary internal function of the family; that is, facilitating the psychological health of family members (Lewis et al., 1976).

Highly competent families are characterized by an effective, intense parental alliance (in two-parent families), clear boundaries between the parents and children, clear communication, good problem-solving capabilities, expression of a broad range of affects, well developed empathic capacities, flexible approaches to change, and the capacity to deal openly with loss.*

The competent but pained family, intermediate between highly competent and clearly dysfunctional families, is characterized by an underlying parental conflict regarding the levels of closeness and intimacy in the marital relationship. The parental conflict is, however, limited to that alliance, and the children are not significantly involved. The focus of such families is the children and the parents share a joint commitment to them. Communication is clear, but less spontaneous than in highly competent families. The expression of affect is somewhat limited, and empathy is compro-

*More detailed descriptions of these and other family types are reported elsewhere (Lewis et al. 1976; Lewis, 1979).

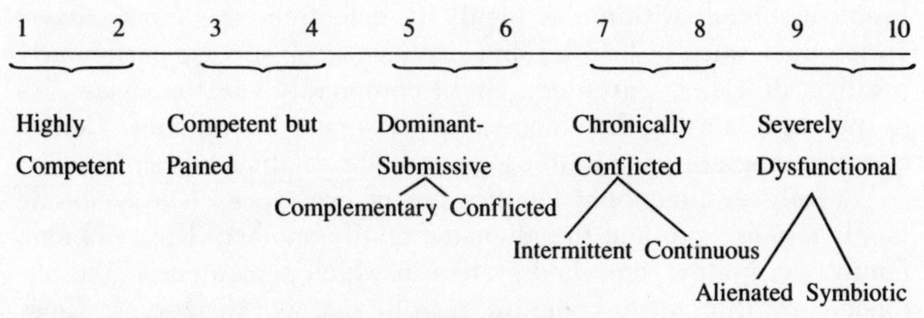

Figure 8.1 The Continuum of Family Competence

mised. There is less overall closeness and often a sense of considerable distance between family members.

The phrase "competent but pained" was selected as a result of our earlier descriptive work with families containing adolescent children in which we found that the levels of psychological health in the children were equivalent to those found in highly competent families: the pain was restricted to the parental alliance (Gossett, Lewis, & Barnhart, 1983; Lewis et al., 1976).

Dominant-submissive families that are complementary demonstrate a clear power differential between the parents that appears acceptable to both parents. These families range from those with a relatively small power difference to those with a vast difference in which the less powerful parent is relegated to (and accepts) a more childlike role. These families, clearly authoritarian, are relatively rigid systems: clear, but without flexibility or spontaneity. Expression of affect is blunted, considerable distance between members is the rule, and dealing with change is a struggle. Dominant-submissive families are seen as dysfunctional systems because there is an increased probability of the development of psychiatric syndromes in the submissive spouse or in the children.

Chronically conflicted families are characterized by a parental relationship in which discord is severe and chronic. The parents' conflict involves their inability to agree on the basic dimensions of their relationship structure. Much of what transpires in such relationship systems can be understood as strategic attempts to gain control of the relationship. The children are often caught up in the parental conflict. Cross-generational alliances may be either transitory or stable. Communication may be difficult to follow.

Problem-solving within the family is defective, the family mood varies from wary to hostile, and expression of affects, particularly positive affects, is curtailed. These chronically conflicted families do not facilitate the development of closeness or intimacy. Consequently, they are poor training grounds for relationship skills.

Severely dysfunctional families are of two types. The symbiotic family is one type, and the alienated family another. The symbiotic family is a highly enmeshed system in which separateness and autonomy are difficult to come by, save by massive distancing. Communication is often difficult to understand, foci of attention are not often shared, and problem-solving is ineffective. There is a primitive, fused quality to the interactions within such families. Individual indentities are blurred and invasive, "what you really feel or think" qualities are often present.

Alienated families are characterized by a profound lack of connectedness between family members. Communication is often stilted, formal, and devoid of affect. Each family member is very separate from other members of the family, and often this distancing is seen in all of his or her relationships. There is little that is warm or nurturing about these family systems.

The family types on the Continuum are seen as ranging from clear and flexible systems (highly competent) to clear but rigid systems (dominant-submissive and chronically conflicted) to chaotic or totally disengaged systems (symbiotic and alienated). As one moves across the Continuum, the conditions for facilitating health of family members decline.

The structure of each type of family may change under changed developmental circumstances or significant stress. Stress, for example, may result in change in a family's interactional structure, and if the change is in the direction of dysfunction (and it often is), the structure can be noted to move from flexible to rigid to chaotic. If the stress results in improved functioning (as is sometimes so), the change is from chaos to rigid to flexible.

These system structures or family types describe families from an interactional perspective. They can best be understood as macroscopic descriptions. The clinician wishing to understand the marriage or family system in which his or her patient is a participant needs also a more microscopic view of that system and the patient's role in the system. A small number of general constructs address the relationship of an individual's psychiatric disturbance to his or her interpersonal systems.

First, marital and family interactions play a variable role in psychopathology. For some patients they are crucial factors in the development or course of a syndrome; for others they play but a modest role.

Second, the relationship between the patient's syndrome and marital and family variables is circular. The marital and family variables influence the syndrome, and the syndrome influences the marital or family relationship system.

A third construct is that there is a significant relationship between the competence of the marital and family system and the probability of family members' psychological health. The more impaired the overall functioning of the family, the greater the probability of psychiatric syndromes in one or more family members (Lewis et al., 1976). In addition, more severely dysfunctional families are apt to facilitate more severe psychiatric syndromes. Although psychologically healthy children may emerge from severely dysfunctional families, and well functioning families may contain individuals with psychiatric disturbances, these are relatively uncommon. The key is the use of the word "probability" to indicate a significant relationship between family competence and individual functioning.

A fourth construct is that, in patients for whom marital and family variables are known to play a crucial role, the patient's symptoms are understood to have communicative meaning. Further, in the case of a particular patient, symptoms may have multiple meanings. Examples of common meanings of an individual's symptoms within a relationship system are: (1) an expression of system dysfunction such as, for example, the inability of the system to modify its interactional structure to meet the demands of a changed context; (2) an expression of the symptomatic person's attempt to exert power in order to change some characteristic of the relationship system—for example, depressive symptoms as an attempt to negate the power of a spouse; (3) an expression designed to avoid or minimize the effects of a more serious and potentially disruptive system problem—for example, an adolescent's behavioral symptoms diverting attention from severe parental marital conflict; (4) an expression of a system projection, as in scapegoating; or (5) an expression of the system's inability to facilitate or tolerate the separation of a system member—for example, in some instances of school phobia.

A fifth construct involving the relationship between marital and family systems and individual psychopathology is that similar relationship patterns and psychiatric syndromes are often found in several generations of the same family. Psychopathology that is transgenerational is not necessarily hereditary. I have known, for example, a family in which for three generations strong women have married charming, dependent men who, after some years of marriage, developed alcoholism. The key to understanding this process was the passage through three generations of a specific relationship pattern, a pattern that facilitates alcoholism.

Although it is not always possible for the clinician to do a marital and/or family interview as part of the assessment of the individual patient, whenever such procedures are possible they often provide data crucial for the understanding of the patient's psychopathology. Clinicians who do such interviews more or less routinely know how often direct observational data are at variance with the patient's self-report regarding his or her marriage and family. I do not wish to discount the importance of the patient's subjective reality, only to emphasize the importance of the clinician's having another perspective. Indeed, a natural starting place for such interviews is the clinician's request that each family member describe the "problem" from his or her perspective.

It takes most students some time to become comfortable interviewing couples or families. Although the residents in our training program have seminars devoted specifically to these issues, I emphasize several aspects of interviewing techniques. The first is that of facilitating interaction. Because interactional data are to be considered, the interviewer needs to use techniques that promote discussion between the spouses or family members. Although much can be gained by soliciting each participant's opinions and beliefs, observing the participants interacting with each other reveals repetitive patterns.

I suggest to beginners that rather than trying to take in everything about the couple or family at one time, it is less confusing to make more limited serial observations. Usually, the distribution of power within the system can be easily ascertained by focusing on who takes the lead in the discussion, who asks questions, who interrupts whom, and who changes the subject. After focusing on power, the interviewer might selectively observe the affective subsystem: what feelings are expressed, who expresses which feelings, how are feelings modulated, what is the level of empathic responsiveness, and how clearly are various affects expressed. The interviewer might then shift to another perspective; and then, perhaps, back to observations about power.

It is helpful to introduce beginning students to rating scales that quantify these dimensions. The Beavers-Timberlawn Family Rating Scales* represent a reasonable starting point for beginners; they can be used with either couples or families and have been used broadly.

When the interviewer is a facilitator and observer, his or her role is relatively inactive. One of the many pleasures of seeing couples and families (in addition to individuals), however, is that it is a much more active type of clinical work than many forms of therapy with individuals. I like the balance provided by the mixture of individual, marital, and family therapies.

A valuable guide for interviewing a couple or family has been provided by Tomm (1987, 1987a, 1988). I introduce residents to his work, although specific training exercises are beyond the scope of this seminar. Tomm believes that the types of questions used by the interviewer have major implications, that some questions are therapeutic and others are countertherapeutic. By dividing the therapist's assumptions into lineal and circular categories and dividing his or her intentions into orienting and influencing categories, Tomm provides a framework for understanding four major groups of questions.

Lineal questions are based on lineal assumptions and are primarily orienting in nature. They are often useful in joining the interpersonal system by appearing to follow system members' typical lineal views of problems. They may, however, convey judgmental attitudes and are inherently conservative (in that they do not invite change) and reductionistic (implying a search for *the* cause).

Circular questions are also orienting in purpose but are more exploratory than lineal questions. The guiding conceptual perspec-

*See Appendix.

tive is interactional, and the search is for repetitive patterns that connect perceptions and events. The aim is to increase family members' awareness of the circularity in their patterns of interaction.

Strategic questions are lineal and influencing in nature: Their intent is primarily corrective. The therapist behaves like a teacher who is trying to change something that is clearly wrong with the system. In doing so, the therapist is imposing what he or she believes to be the *correct* solution, and this potentially coercive impact can lead to countertherapeutic effects.

Reflexive questions are circular and influencing in nature and are meant to facilitate the mobilization of family members' own problem-solving capacities. They are reflexive in the sense that they are meant to trigger the reflections of family members regarding their current operations and consequently to have a generative effect on the system.

Tomm states that differentiating these four categories of questions does not depend on their structure or content, but rather on the interviewer's intentions and assumptions. Therefore, the emotional tone used in asking the questions can be the decisive indicator of what type of question is being asked. Although each type of question has utility, it is clear that Tomm believes that circular and reflexive questions flow directly from a systems approach to psychopathology and, hence, are much more apt to have a positive therapeutic impact than are lineal and strategic questions.

Although there is much, much more to teaching residents the basic skills involved in interviewing couples and families, they are introduced to the subject during the seminar because of my belief that a clinical formulation is incomplete unless the patient can be interviewed with spouse or family.

For couples or families referred as such, I encourage several conjoint or family interviews and one individual interview with each family member. These interviews are designated clearly as assessment, although they often have considerable therapeutic impact. They are termed assessment interviews because I am uncertain at the beginning as to what I will recommend. Often the recommendation is for marital or family therapy, but at times some other form of treatment such as individual psychotherapy or psychotropic medication may be the primary recommendation.

TABLE 8.1
Format for Observations
of a Marital System

I. Interactional Structure
  A. Highly Competent
  B. Competent but Pained
  C. Dominant-submissive
  D. Chronically Conflicted
  E. Severely Dysfunctional
    1. Symbiotic
    2. Alienated
II. Key Structural Variables
  A. Level of Connectedness
    1. Closeness
    2. Intimacy
    3. Commitment
  B. Level of Separateness/Autonomy
  C. Power Distribution
III. Levels of Satisfaction
IV. Developmental Factors
V. Generational Issues
VI. External Stress

Often this type of detailed marital or family assessment is not feasible for the individual patient, and the clinician must base his or her observations on a single marital or family interview.

There are several special procedures that I have found helpful. One is to have each spouse or family member complete the Rokeach Value Survey independently in order that I understand something of their agreements and differences regarding basic life goals (Rokeach, 1973). A second special procedure involves video-taping a marital or family interactional task to use in playing back to the couple or family.

Residents also need a brief and relatively simple format for organizing their observations about the marital and family systems. Table 8.1 presents such a format for the patient's marital system.

1. *The Interactional Structure.* The students are asked to arrive at a macroscopic picture of the marital system interactional structure using the marital types derived from the research at the Timberlawn Psychiatric Research Foundation (Lewis, 1989; Lewis et al., 1976; Lewis & Looney, 1983). They are instructed that, although

the correlations between the individual psychological health of spouses and the competence of their marital systems are positive and statistically significant, there is considerable variation (Lewis, 1989). Two psychologically healthy spouses can, for example, have a dysfunctional relationship. One healthy spouse and one with less than average psychological health can have a highly competent relationship. Thus, knowledge of the individual patient's psychiatric syndrome does not inform as to the quality of his or her marriage.

2. *Key Structural Variables.* The importance of these variables is based on the assumption that they represent issues that must be dealt with in the process of establishing a relationship with a future. In concert, they concern a central issue in the establishment of a relationship: the balance of connectedness and separateness.

Closeness is defined as the amount of sharing in the relationship. How much the spouses share values, activities, interests, friends, and satisfaction in their sexual relationship is a key issue.

Intimacy refers to the reciprocal exchange of private thoughts, feelings, fears, hopes, wishes, and fantasies about which they feel a sense of vulnerability.

Commitment is defined as both the intensity and loyalty to the relationship in comparison to each spouse's other relationships. A highly committed relationship is one in which both spouses experience their relationship as the single most important relationship in their lives.

Separateness/autonomy refers to the degree to which the relationship structure facilitates separate interests, activities, values, and friends, as well as the capacity of each spouse to function independently.

Power is defined as interpersonal influence. The spouses must decide who is to make decisions, how disagreements are to be resolved, and to what extent negotiation and compromise are to be sought.

3. *Levels of Satisfaction.* The relationship between the interactional structure of a relationship and the participants' satisfaction with the relationship is not strong. Although spouses in highly competent marriages are more apt to be satisfied than spouses in dysfunctional relationships, there are many clearly dysfunctional relationships in which both spouses are satisfied because many factors

other than the relationship structure influence the spouses' level of satisfaction (Lewis, 1989).

4. *Developmental Factors*. Relationship systems traverse a number of developmental stages much as individuals do. Transitional periods as, for example, the assumption of parenthood or the last child's leaving home are often times of stress within the system. Although our research suggests that the outcome of such transitions varies according to the nature of the system's interactional structure prior to the stress, clinical work often reveals that transitions are times in which one or more participants may develop symptoms (Lewis, 1989). It is important, therefore, that clinicians know which developmental period the patient's marital system is in and whether the patient's symptoms are related to a stressful transitional stage.

5. *Generational Issues*. Many family theorists emphasize the connections between marital and family processes at two or more generational levels. A child, for example, may develop symptoms at a time when his or her parents are struggling with a potentially relationship-ending conflict and the grandparents are caught up in issues having to do with retirement. It thus becomes important for the clinician to understand something of what the current issues may be at several generational levels.

6. *External Stress*. Unemployment, serious medical illness, death of a parent, and many other severe external stresses may impinge upon the marital relationship. Although these processes may strengthen the relationship at times, often they play a destructive role. Many factors influence the impact of external stress on the marital system, and the clinician's knowledge of these circumstances is often helpful orienting information.

> The patient, a 46-year-old housewife, presented with symptoms of a major depression. As one part of the evaluation of this patient, the clinician interviewed the patient and her husband together.
>
> The patient and her attorney husband have been married for 25 years and have daughters 22 and 18 years old. The marital structure is dominant-submissive; the husband is clearly more powerful than the patient. They have had little

conflict about this structure; it seems to have consciously suited each spouse. The patient has been very dependent, with few areas of autonomous functioning.

There is considerable closeness in the relationship; the spouses agree about and share traditional Christian values. The wife joins her husband in his activities. Their friends are, for the most part, from his legal practice. Their sexual activity is reported to be mutually satisfactory.

Each spouse appears to be highly committed to the marriage, but psychological intimacy is absent. The patient reports that her husband does not deal openly with "private matters," and he agrees.

The patient's autonomy is significantly compromised by her strong dependency on her husband. The husband describes a high level of satisfaction with the marriage. The patient describes much satisfaction, although volunteering that his busyness and emotional remoteness often leave her feeling lonely.

The patient's depressive symptoms began in the summer before her younger daughter was to leave for college. Her relationship with this daughter is described as "very close" and "quite confidential."

The patient's parents are alive, but her father's failing health has led to consideration of his impending need for a nursing home placement. Other than these issues, there is no known source of stress on the patient or the marital system.

Although this marital formulation is brief and at a macroscopic level, it places the development of the patient's major depression in the context of her relationship systems. It raises questions about factors that may play an important role in the etiology of her depression and facilitates the clinician's consideration of a range of treatment modalities.

Table 8.2 provides a format for organizing the clinician's observations about the patient's family. It is particularly useful when the patient is young—a child, an adolescent, or a young adult.

1. *Interactional Structure.* As with the macroscopic assessment of the marital interactional structure, it is helpful to start the family assessment with a judgment regarding the overall level of family

TABLE 8.2
Family Assessment

I. Interactional Structure
II. Parental Marital Formulation
III. Boundaries
  With Outside World
  With Extended Family
  Within Family
IV. Coalitions and Triangles
V. Communication
  Clarity
  Spontaneity
VI. Affect
  Openness
  Range
  Empathy
  Conflict
VII. Developmental Issues
VIII. Generational Issues
IX. Values
X. External Stress

functioning. For this purpose, the students are introduced to the Continuum of Family Competence (See Figure 8.1) derived from the Timberlawn Psychiatric Research Foundation's research done with both well-functioning and dysfunctional families (Lewis 1989; Lewis et al., 1976; Lewis & Looney, 1983). Students can be taught this system of classifying families without difficulty, and it provides a perspective on the macroscopic structure of the family.

This approach to the classification of families based on their competence in facilitating family members' psychological health has been used in a variety of studies and has been found to be a useful perspective. A problem with this approach to family classification, however, is that it has not been validated on single-parent or recombined families.

2. *Parental Marital Formulation.* In two-parent families, the parental relationship is the template for the structure of the family. If the parental relationship is significantly dysfunctional, it is difficult for a family to function with full effectiveness. Our studies of families containing adolescent children demonstrated a significant correlation between independent assessments of the parents'

marital relationship and total family functioning (Lewis et al., 1976). Occasionally, however, parents with underlying conflict can keep the conflict within their relationship and not bring the children into the conflict. This appears to explain the high levels of psychological health in children whose parents have a competent but pained marriage.

The format used in this part of a family assessment is the marital assessment outline displayed in Table 8.1.

3. *Boundaries.* In the sense used here, a boundary represents the processes both of connecting and of separating from others. A family's boundary thus reflects how much involvement the family has with the outside world, with friends, neighbors, and surrounding agencies, and the nature of that involvement. A family with a rigid, relatively impermeable boundary is isolated from the outside world. Many families have considerable involvement with the outside world, but are also able under certain circumstances to separate from that world. Meals, family celebrations, and family rituals are examples of activities that often exclude others. Finally, the boundaries of some families are so permeable that there is little sense of a cohesive family.

The nuclear family must also establish boundaries with the extended family. Ethnic and cultural forces may shape such boundaries, and there is a broad range of possibilities extending from families that are isolated from kin to those that differentiate but slightly between cousins and siblings and parents and grandparents.

Boundaries within the family mark off the parental alliance from the children. Time for parents separate from the children and appropriate privacy for the parental relationship are examples of such boundaries.

Each family facilitates the establishment of a certain level of individual boundaries. In some families, the individual boundaries are so rigidly drawn that family cohesiveness suffers and family life is experienced as very distant. Other families do not facilitate individual interests and activities and provide little help in family members' work to distinguish *me* from *not-me*. Respecting each individual's right to have private feelings and thoughts is an important aspect of boundaries within the family.

4. *Coalitions and Triangles.* Studies of well-functioning families find that a strong parental coalition is central. When the parental coalition is dysfunctional and unsatisfactory, one or both parents may enter into intense coalitions with children. These cross-generational coalitions are often understood as dysfunctional triangles in which the parent-child coalition excludes or opposes the other parent. Many family therapists believe that a child who is caught up in a dysfunctional triangle is at increased risk for the development of psychiatric symptoms.

5. *Communication.* The clinician needs to attend the clarity of the family's communication system. Better functioning families facilitate family members' open expression of thoughts. Severely dysfunctional families often present with chaotic, difficult-to-follow communication in which a listener is mystified as to who thinks what.

Well functioning families are spontaneous in their communication. Humor is frequent. Moderately dysfunctional families present with a more stilted, formal pattern of communication or, in the case of chronically conflicted families, the escalating conflict involves rapid, bitter speech.

6. *Affect.* Family systems evolve characteristic processes regarding the expression of affects and responses to affects. Well-functioning families are open to a wide range of expressiveness; empathic responses within the family are common, and a general tone of freedom with affects pervades. More dysfunctional families are less openly expressive of affects, restrict the number of allowable affects, and are far less likely to exhibit empathic responsiveness. The presence of underlying conflict with associated anger, bitterness, and blaming restricts the expression and range of affects.

7. *Developmental Issues.* As with marital systems, it is often useful for the clinician to know that the patient's family is caught up in a developmental issue. Whether it is the birth of another child, the "letting go" associated with a child's entering school, the adjustments required to assist adolescent children to experiment with increased independence, or later family developmental transitions,

the knowledge that the family is in a transitional period of adjustment to change is helpful. Such periods in family life are often associated with anxiety and uncertainty. Family members are more vulnerable to the development of symptoms during such periods.

8. *Generational Issues.* The issues involving clinically relevant processes at two or more generational levels is the same for the family system as described for the marital system. Many family therapists believe that a transgenerational process can start in one generation and present clinically in a different generation. Thus, concern for the grandparents regarding a problem such as retirement may lead to anxiety or conflict in the parents which, in turn, may play a role in a grandchild's refusing to go to school.

9. *Values.* Families usually have shared values. During adolescence there may be some rebellion against family values, although the family values are usually accepted by the children. The clinician's knowledge of the patient's family values can be important in treatment planning. Of course, the clinician should be aware of the patient's values that differ from his or her own values.

10. *External Stress.* As was noted in the discussion of the marital assessment, external stress in many forms can deleteriously influence a family's structure and functioning. Knowledge that the patient's family is struggling with a major stress from without often provides the clinician with clues regarding the impact of the family on the individual patient.

> The patient, a 19-year-old college student, was referred for evaluation because of declining school performance, heavy drinking, and a series of automobile accidents. As one part of his evaluation, the clinician interviewed the patient's family twice. The interviews included both parents, the patient, and a 17-year-old sister.
>
> The family interviews resulted in the following findings:
>
> The family's interactional structure is chronically conflicted. The parents, married for 20 years, have never been without underlying conflict. Each parent attempts to control the other through lies, distortions, and manipulations. There

is little closeness and no intimacy in their relationship and, as a consequence, little in the way of family cohesiveness.

Family boundaries are weak. Friends and relatives are often in the home and appear to regulate the parent's underlying conflict. The marital relationship does not have a clear boundary as each parent has an important coalition with the child of the opposite sex. The family's basic structure is thus comprised of two triangles. The patient is involved in an overly close relationship with his mother which excludes the father, who is seen as the victimizer.

The family's communication is clear, but full of attacking and blaming. With the exception of anger, most other affects are not expressed. Affection appears to be expressed only in the mother-son and father-daughter relationships.

The patient's departure for college interrupted the family's chronically conflicted but relatively stable state. Mother's loss of her ally left her feeling lonely and an outsider. In response, she called her son frequently and discussed with him her growing wish to divorce his father.

Both sets of grandparents are alive and divorced. All four grandparents visit the family home frequently, often staying for several weeks at a time.

The family's value system reflects both traditional middle class values emphasizing individuality, successful competition, and an economic assessment of individual worth. In addition, the family members appear to believe in God and a life after death, but seldom go to church.

There is no identifiable external stress impinging on the family.

The clinician concluded that family issues played a prominent role in the patient's symptomatology. Relevant factors included the chronically conflicted parental relationship and the patient's enmeshed relationship with his mother. The patient's departure to college was seen as the precipitating event. The patient's symptoms were understood from the family perspective to have the following meanings:

1. An expression of family dysfunction and a response to change in family structure that was occasioned by his leaving home.

2. An attempt to divert attention from the parental conflict and end the threat of divorce.
3. An attempt to make further participation at college impossible, thereby necessitating his return home and the restabilization of the family structure.

Although these marital and family perspectives on psychopathology are but a beginning, it is important to introduce them early in the students' learning process. It is important also to emphasize that such perspectives are complementary to biologic and intrapsychic perspectives. By using all three perspectives, the student is more likely to understand the patient's uniqueness and to use a broad range of treatment interventions.

CHAPTER 9

Other Issues

SELECTING RESIDENTS

For most residents the seminar experience is a very positive part of the initial stage of training. I am not sure that such a favorable response would be found in all training programs. To a considerable extent, this surmise is based on the observation that training programs do not share a universal set of criteria for the selection of resident applicants. In some programs, criteria may not be articulated clearly, and in others the faculty may disagree about which criteria are most important. It also seems apparent that some programs publicly endorse one set of criteria, but rank applicants on the basis of other criteria.

Each department of psychiatry or psychiatric hospital has a prevailing ideology from which the institution's beliefs about the profession's core distinctions are derived. Inevitably, the curriculum of the institution's training program reflects these ideas. Thus, the emphasis given to biological, psychological, or social system variables is clearly related to these shared value judgments.

The current neuroscience bandwagon has modified the ideas of many, if not all, training programs about the profession's core distinctions. This is inevitable. The issue, however, is the degree to which the increased emphasis on biological determinants has replaced prior emphases on psychological and social system determinants. Although for all departments money talks, its voice is louder in some than in others. If much of the department's funding comes from research grants, and as it is biological research that is predominantly supported by granting agencies, biological researchers will be increasingly powerful faculty members. The attitudes of such biologically oriented persons toward psychotherapy training, although variable, is not, in general, strongly positive.

Some years ago, my eldest child decided to pursue psychiatric training. As he prepared to make "the trip" both to be interviewed by the faculty of a large number of training programs and to get some sense of each institution's values and curriculum, he asked me what to look for. My response was to attend as much as any other factor how residents at each program said a "good" resident was defined. Although his trip was over 10 years ago, it was surprising how many residents even then defined a successful resident in their program as one who participated in and coauthored reports of double-blind drug studies. The acquisition of broad clinical skills with a major emphasis on psychotherapy was clearly the most important characteristic of a "good" resident in some, but far from all, training programs.

One of the more intriguing of the current dialectics about preferred applicant attributes—intellectual brightness versus human qualities—has, I believe, been in place throughout my career. Although outstanding candidates are seen as having both attributes, more often candidates may be perceived as stronger in one dimension than in the other.

Another interesting issue is the presence of clear psychopathology, its quality and intensity. What, for example, will be a training faculty's response to an applicant with a history of adolescent drug abuse, depression, or psychosis? Does the presence of such a history with clear evidence of reasonably mature current functioning serve as a positive or negative variable in the selection process? How will various training programs respond to the application of the brilliant, but clearly obsessional, candidate? These are intriguing questions about which there are little systematic empirical data.

Perhaps in the best of all worlds a good "fit" occurs. Residents with attributes valued by a particular program will be drawn to that program and probably be ranked highly by the program. That such a fortuitous circumstance occurs only some of the time is clear to those of us who visit many programs and talk with many residents.

I became part of the medical house staff at Peter Bent Brigham Hospital in Boston in the early 1950s and recall vividly my surprise at finding in my new colleagues broad interest in psychiatry. Nothing in my Texas medical school experience had prepared me for that level of interest. In a setting that then had difficulty in constraining its pride in attracting "the best and the brightest," psychiatry was seen as more than a merely acceptable option.

Certainly, a lot of the difference in attitude between Dallas and Boston in 1951 was the different intellectual climate of the two cities. The institutions themselves were vastly different. My medical school in Dallas was young, small, and committed to producing general practitioners for rural parts of the state. There was no push to specialize, and no individuals or institutions of sufficient prominence to serve as inducements for careers in psychiatry. Psychoanalysis had little apparent impact on the small group of psychiatrists in the community who, for the most part were biologically oriented and, in some instances, anti-analytic in their perspectives and behaviors.

On the other hand, Peter Bent Brigham in the early 1950s often selected its chiefs of medicine and surgery from very young candidates hardly out of residency training. The hospital, closely tied to Harvard Medical School, seemed so certain of its excellence that it could gamble on very young persons as future leaders.

Additionally, Boston seemed to experience itself as the intellectual capitol of America and, as such, had embraced psychoanalysis—indeed, well before the early 50s it was part of the central fabric of Boston intellectualism. The Boston Psychopathic Hospital was psychoanalytically oriented and played a major role in the intellectual ferment of Boston medicine of that era.

Bright young medical students in Boston were attracted to psychiatry in part because that's where the intellectual action was and, in part, because there was no absence of stimulating role models.

Over half of Brigham's medical house officers were from Harvard Medical School and, as a group, they seemed different from my Texas medical school classmates. There was no hint of "the good ole boy" predominant character structure of that era's Texas medical schools.

My roommates at the Brigham were in some ways representative. One was the newly appointed chairman of a basic science department of a medical school in a different city while he worked 36 hours on and 12 hours off as an intern at Brigham. The other was torn between a career in what was then called "protein chemistry" and a commitment to a psychoanalytic career. His choices seemed so radically different that I had trouble comprehending. "Jerry," he would say, "think of it as a contest between the aesthetics of interpretation and the purity of certain amino acid structures." At the time I only vaguely understood what he was talking about, but it seemed quite clear that some very bright young people found psychiatry to be an exciting possibility.

All of this is by way of introducing my experience in interviewing and participating in the selection of psychiatric residents from a large pool of applicants. Over the 25 years of interviewing applicants, a reasonable estimate is that I have interviewed 600 young men and women. From the early days, I have asked each applicant the same two questions: "How does it feel to be you?" and "How do you relate what it feels like to be you to the family you grew up in?"

I use these questions because our faculty has been clear in its values that the program's basic goal is to graduate residents with a broadly-based, multivariable model of psychopathology with emphasis on biological, psychodynamic, and social system determinants of behavior, a psychiatrist who is distinguished by his or her ability to help others explore their experiences. Thus, the facilitation of psychotherapeutic competence is central to the mission of the training program, and it is important to transform the values into curriculum time, case loads, and supervisory hours that affirm those beliefs.

Thus, the challenge has been to select residents from a large pool of applicants who are best suited for a training program that emphasizes the development of psychotherapeutic competence.* From this perspective, my two questions try to tap into a candidate's capacity to articulate a narrative about his or her life. The

*Each candidate is interviewed by both faculty and current residents, each of whom records his or her observations independently on rating scales. The candidate's interview scores are averaged and, along with the Dean's letter (and clerkship evaluation if done at Timberlawn), form the basis for assigning a ranking. (The rating scales are replicated in the Appendix).

features of a candidate's narrative that I have found particularly useful in rating the candidate's application are its level of complexity and richness. Although some candidates have experienced more dramatic episodes, unfortunate life events to be overcome, dysfunctional family systems, and the like, what is important is the nature of the linkages established between events, persons, and experiences. I view a candidate's ability to link such pieces into themes or organizing metaphors as a positive selection factor. My premise is that narrative construction is an important part of the psychotherapeutic process, and candidates with advanced narrative capabilities are preferable to those who demonstrate little such ability.

A second predictive feature is a candidate's relationship to his or her narrative. Whether a particular candidate's narrative is rich and complex or less so, I am interested in whether a candidate appears to reexperience some of the affects associated with the narrative. Some degree of experiencing affects is seen as preferable to distant and matter-of-fact accounting. My belief is that the capacity to be empathic with one's self—particularly with one's earlier life experiences—is often reflected in the affect experienced while telling one's story. Because empathy is an important psychotherapeutic skill, I view candidates who demonstrate empathy with themselves as preferable to those who do not.

Content is perhaps the most difficult aspect of the candidate's narrative to evaluate. How much emphasis should be placed on overcoming adversity in contrast to a story of relatively smooth development? Is a history of intense and broad interest in relationships always a positive factor, or does it sometimes reflect object hunger? What does it mean if there is no evidence of a close, confiding, intimate relationship in the candidate's story?

Levinson (1978) suggests that individuals may traverse young adulthood as either searchers or trackers. Most of the candidates for our program are trackers who made a decision to pursue a medical education during adolescence and stayed successfully "on track." Applying for a residency is a piece of this continuous process or track. A sizeable minority of our applicants are clearly searchers. As a group they are a little older; they have explored other areas before deciding to pursue medical training. Some have been teachers, psychiatric technicians, poets, nurses, television reporters, journalists, or other occupations. Sometimes their interest in medical school developed because they became interested specifically in psychiatry. Many have been involved in outpatient psychotherapy during earlier periods of their lives.

Although judging a candidate would not be done solely on the basis of these two styles of traversing young adulthood, how should they be valued? More often than not, I see searchingness as positive, particularly if the candidate's experiences outside the usual track have appeared to enrich his or her life.

A related feature involves those candidates who have been practicing or in training for another specialty. Once again, it is not the fact that one is an ophthalmologist or surgeon and wishes to pursue psychiatric training, but rather the way in which the change is embedded in the candidate's narrative and is understandable from the perspective of that narrative's major theme or themes.

Another set of observations involves the subtle suggestion of a heroic theme in the candidate's narrative. I was taught by supervisors that any suggestion of the need to rescue others was neurotic and needed to be monitored closely in one's psychotherapeutic work. I have come to have a different perspective. Although the need to act heroically in rescuing others can be excessive in its urgency, to be totally without such motivation would be to be deficient in humanity. Because the need to be helpful is related to altruism, a higher level mental operation often associated with mature character formation, its absence in a candidate's narrative would concern me (Vaillant, 1971).

I am interested in both the extent to which candidates commence their narrative in the family of origin (as they are asked to do), and whether they incorporate their role in the family as a significant theme. Many candidates describe families of origin with strong parental coalitions, high levels of cohesiveness, and open communication, and they have always felt support for pursuing their own interests, including medical school and psychiatric training—all characteristics of well-functioning families.

Other candidates describe clearly dysfunctional families of origin. In some such families there is a chronically ill family member—a depressed mother, an alcoholic father, or a schizophrenic sibling. Some candidates acknowledge that their early role in the family was that of the helper or, perhaps, the go-between for a conflicted parental relationship. Other candidates appear not to have connected these family circumstances to the current aspects of their narratives.

No single element is decisive in the candidate's description of his or her family of origin or his or her role within the family. Many outstanding residents come from clearly dysfunctional families in which they played an early therapist-like role. Others come from well-functioning families in which such family roles were not necessary. The manner in which the candidate is able to include family experiences in his or her narrative is important. Whether it is the struggle to emerge with as few scars as possible from a very dysfunctional family or the continuation of the self-determination encouraged in a well-functioning family, candidates who incorporate those experiences in their narratives, explore possible connections between present and past, and are open to their own investigative probes are the strongest candidates.

A fourth perspective from which I evaluate a candidate's narrative is that of an encounter with an apparently powerful senior person. If the candidate is strongly invested in acceptance in our program, the interview has the additional component of an attempt to please, to present one's self in as favorable a light as possible. Those who comment on the here-and-now of the context or their reaction to it reflect an ability to be both sensitive to the immediate interpersonal context and to be open about their observations. Because this type of contextual sensitivity is an important attribute of the psychotherapist, when a candidate makes such observations during the interview I consider it very positively. Regardless, the candidate's narrative can be understood as an attempt to deal with, win over, neutralize, or hide from a powerful parental surrogate. As such, this perspective has important implications for the selection process.

For many years each candidate who passed the training director's initial scrutiny was interviewed by a minimum of six faculty members and six residents over a two-day period. An increased number of candidates has made the two-day evaluation logistically impossible, so now one day of interviews has to suffice. Many candidates assess the program more thoroughly and, hoping to improve their chances, take a clerkship at the hospital for four to six weeks. Generally, faculty and residents agree on various candidates' strengths, but often one faculty interviewer disagrees sharply with his or her colleagues about a particular candidate. Although, this is most likely the result of a less-than-adequate interaction between interviewer and candidate, it is sometimes surprising how tenaciously the interviewer persists in ascribing his or her deviant rating of a candidate as a reflection of the candidate's characteristics only.

The opportunity to interview 600 or so bright young medical students and physicians has been a rewarding experience—particularly to be able to compare my impressions with those of a small group of colleagues who have served on the Residency Committee for many years. I am told that I am an "easy" rater, often inclined to invoke a "high risk-high gain" perspective on candidates with unusual stories. Perhaps those observations are valid, but when we have selected a resident who doesn't do well in our program, it often involves being dazzled by an applicant's brilliance and intellectualism and somehow overlooking his or her relationship deficits.

SELF-MONITORING

In these days of intense and often hostile monitoring of psychiatric treatment by representatives of managed health care and health insurance, it may seem too much to suggest that psychotherapists need to monitor their own work and that training in doing so needs to begin during residency training. Such is so, however, and has been so since psychotherapy became a formal discipline. Competent psychotherapists monitor their own work, sometimes adding supervision, consultation, or small-group activity. Each of these processes is valuable. What is important is that psychotherapists have a systematic format within which to examine their own work.

In 1979 I published a suggested format for self-monitoring (Lewis, 1979a). To review briefly, the central idea of that paper was that there are two very different types of psychotherapeutic work: that which occurs at considerable emotional distance from the patient (the "objective" position) and that which is accomplished when one is working very close to the patient's experience (the empathic position). Both types of psychotherapeutic work need monitoring.

As described in earlier chapters, the objective or detached work of psychotherapy includes a variety of the therapist's activities, all of which tend to focus on the patient's psychopathology. The goals of such activities are to understand and to influence. Included in these activities are following the patient's associations, recognizing nonverbal signals, indentifying mechanisms of defense, and all the other observational processes involved in psychotherapy. A second group of detached processes are the mental operations used by the therapist to understand the patient's dilemma. The use of these operations permits the therapist to bring together his or her observations and inferences in the form of a clinical formulation. Finally, there are the basic interventions: facilitation of the patient's self-exploration, confrontation, clarifications, and interpretation.

The empathic position is centrally involved with sensitivity to affective messages and the empathic process. The empathic process begins with the recognition of the patient's affective messages (cognitive empathy) and then arouses a similar affective state in the therapist. The affective arousal of the therapist may take the form of an affectively ladened fantasy or memory or even a relatively content-free affective state. I suggested further that the empathic process may appear to occur spontaneously and effortlessly in the therapist, or it may be induced deliberately by the therapist when the therapist attempts to "let go" of his or her customary detachment and enter the patient's experience without any sense of where he or she is going.

In the 1979 paper I suggested that both the detached and empathic work of psychotherapy need to be monitored by the therapist by systematically and periodically reviewing three important dimensions of the psychotherapeutic process of each of his or her patients—the clinical formulation, affective arousal, and distribution of power within the psychotherapeutic relationship.

The detached work of psychotherapy can be monitored by focusing one's attention on the broadly based clinical formulation arrived at after the first several interviews. Further, the psychotherapist can periodically ask himself or herself, *"How do I understand the patient's dilemma at this time?"* It is valuable to write out such a formulation in order to compare it with the initial formulation. If there are no changes in one's formulation, it is likely that one has not been doing all the work necessary from the detached position. Clinical formulations should change as deeper levels of understanding the patient occur as a result of the therapist's acquisition of more data.

The failure to change one's understanding of the patient often reflects premature closure brought about by countertransference processes. One may, for example, simply stop trying to deepen one's understanding if doing so might lead to painful aspects of the therapist's life. Another frequent countertransference theme involved when the formulation hasn't changed occurs when the process of psychotherapy has come to be disproportionately empathic; the psychotherapist is unable to disengage himself or herself sufficiently to have arrived at a deeper or broader understanding.

In addition to the absence of change in one's clinical formulation, a common marker of problems in the detached work of therapy is a thin, inadequate formulation, often consisting of but a few broad generalizations regarding the patient's psychopathology. Although beginning residents often present such sparse formulations, they frequently do not know any better. What I refer to here, however, is not lack of knowledge; rather, it is the circumstance wherein one can articulate clearly adequate formulations regarding all but one of one's patients. Factors that lead to an inadequate formulation with a particular patient are often countertransference processes at work, processes that restrict our knowing, often based on the wish to avoid pain.

The empathic processes of psychotherapy also need monitoring and may be even more crucial than monitoring the detached processes. The question to ask one's self periodically is, *"How long since I have been aroused affectively in my work with this patient?"* If the answer is that one has never shared the patient's emotional experience or if it has been several or more interviews since it has occurred, the probability is that one is not sufficiently attending the empathic component of treatment. Even though there are many reasons for maintaining considerable distance from the patient's experience, our concern here is with the processes of countertransference, whether they are operating and what one is defending against.

A third marker for monitoring one's own psychotherapeutic work is a periodic assessment of the distribution of power within the psychotherapeutic relationship. It is interesting how often the data reveal that one is not collaborating with the patient, but is behaving more powerfully in the interviews than one intended. If this is so, possibly countertransference processes are operating that need self-exploration.

The periodic examination of the clinical formulation, episodes of affective arousal, and the distribution of power within the psychotherapeutic relationship provide the therapist with three very different approaches to monitoring his or her own work. My experience with these markers suggests that remembering to use them is often associated with psychotherapy that is going well, but on many occasions one of the three markers may suggest the operation of countertransference. Forgetting to use them often signals the operation of destructive unconscious processes within the therapist. I will, for example, "suddenly" remember that I have not asked myself the marker questions for several months or longer in regard to a particular patient. This occurs even though I am using this monitoring system in regard to my work with other patients. Almost always, the omission is significant and discovering what it is all about usually leads to progress in the psychotherapy of that patient.

The residents are introduced to this system of monitoring their own work during the latter stages of the seminar. The beginning seminar for beginning residents does not review psychotherapy interviews, in contrast to an advanced seminar with third and fourth year residents in which videotaped psychotherapy sessions from their own work with patients are reviewed. Rather, the focus is on exploratory interviews with patients the beginning residents do not know. Even in such one-time experiences, the periodic review of the three markers (along with other aspects of the interview) can be useful. *"How does this information change your initial ideas about the clinical formulation?" "Does it seem to you that you've consistently been at some distance from the patient's experience?"* or *"Can you see how you've loosened up since the opening phase of the interview, how much less controlling and powerful you seem to be?"* These questions help focus the resident's attention on important aspects of the progress of a single interview.

My hope is that introducing the residents to the process of self-monitoring early in their training augments the possibility that such a process will become routine for them. I do not wish to suggest that the three markers I use are necessarily the best ones for everyone. It may well be that part of the process of becoming a psychotherapist is coming to grips with the use of markers that are more specific to one's propensities for certain forms of countertransference behaviors. The use of the clinical formulation, affective arousal, and power in the relationship may work best for me because they tap into processes specific for me. The important point for residents is to incorporate some way of monitoring their work into their developing sense of a psychotherapeutic self.

THE CONTINUING MATURATION OF THE PSYCHOTHERAPIST

The distinction between learning and becoming is important in many aspects of life. A person could know, for example, a great deal about parenting and not be a good parent. I believe it is possible to know a lot about psychotherapy without necessarily being a competent therapist. As important as I believe it to be, I am not certain that we understand very much about this distinction. Clearly it has something to do with the way in which being a psychotherapist is experienced. Being a psychotherapist becomes a central part of the way one experiences one's self. It has much less to do with what one says to people than with the way one listens and the way one thinks about causality. Psychotherapists listen for connections, that which is not explicit. They are sensitive to the paralinguistic aspects of communications in everyday life. They note subtle expressions of affect, those often momentary expressions that pass so quickly a less perceptive person might question whether they were really present.

Psychotherapists come to understand that what appears simple is suspect; that almost everything is the result of multiple causes. Psychotherapists also move toward a constructivist perspective of reality. Having spent so much time listening to patients' subjective and often very idiosyncratic views of their worlds, the listener comes increasingly to question the idea of a fixed, absolute, objective reality. As a consequence, psychotherapists become more respectful of other persons' subjective realities.

Thus, a psychotherapist comes to be, over time, that which he or she does. Having said this, I hasten to add that psychotherapists do not "do" psychotherapy in the supermarket, at church, or in bed. They do not behave psychotherapeutically in contexts where such behavior is inappropriate. Rather, they deliberately suspend the operation of what has become a real part of themselves. Thus, they may cry in the theater, scream at the arrogance of the opposing quarterback, and feel hurt when their names are forgotten. It is, I believe, much like being a parent. There are many contexts in which that very real part of one's self is put aside.

Perhaps another way of looking at the distinction is from the perspective of naturalness. As one becomes a psychotherapist, an increasing amount of one's psychotherapeutic work flows spontaneously, without the labored "*What do I do next*" experience so common during earlier years. This is not to say that a mature psychotherapist is not thoughtful, nor that planned, deliberate interventions are never used. Rather, they diminish in frequency, and one comes to rely on the natural flow of what comes spontaneously to mind.

This change in how the psychotherapist experiences himself or herself while working is important for beginning students to understand. Their initial efforts are of necessity terribly labored. "*What do I do next?*" fills their minds and interferes with their work. There is no escaping it, but the situation will get better. Often I use the simple analogy of going to a tennis coach for help with a poor serve. The coach may say, "Stop doing what you're doing and start over," perhaps beginning with even the most basic of things like how you hold the racket and toss the ball. Doing what the tennis pro has suggested seems foreign, artificial, not you. It is only with practice—doing it repeatedly—that it begins to be natural, to become *your* serve, a part of you.

We have no systematic longitudinal data on the process of becoming a psychotherapist. Therefore, we do not know the influences that may play a crucial role, nor whether the factors crucial for each of us are different. I believe, however, that three common factors may operate to facilitate becoming a psychotherapist.

The first factor is one's relationship to one's work and, in particular, how thoughtful one is about it. I believe most psychotherapists are intrigued and curious, continuing to think about sessions after they are over. I do not mean preoccupation with their work, but rather that the work continues to fascinate and delight them, and each day's work is usually anticipated with pleasure.

I don't know why the work of psychotherapy retains its vitality for some and not for others. Almost certainly it is the result of many factors, and I often think that the seeds must be planted early in training.

Second, I believe that professionals who have evolved relationship systems that provide them with a secure sense of connectedness with closeness and intimacy have greater freedom to invest in the work of psychotherapy because fewer of their own basic needs have to be met in their work. Having said that, I acknowledge immediately that many of the outstanding contributors to the development of psychotherapy clearly did not have good marriages, cohesive families, or supportive friendship networks. I am also aware that many practicing psychotherapists are not strangers to all sorts of relationship difficulties. I continue to believe, nevertheless, that those with effective relationships are more likely to retain an enthusiastic attitude toward their work.

I believe that this correlation is not simply a matter of psychological health, although it clearly plays a role. I say this because of the data that suggest that relationship competence is more than a matter of just the participants' levels of individual psychological health (Lewis, 1989).

It is interesting to ask colleagues who have experienced a dysfunctional first marriage and are now in a much more effective second marriage what differences, if any, they've noted. Without exception, those I've asked have stated that the differences in their work are clear and significant to them. Often, they use the concept of freedom as part of their descriptions. *"I'm much more free to do the work of psychotherapy and find it much more interesting"* is a not uncommon statement. Depression is another frequently used descriptor. *"I think I was depressed and apathetic for years (during the first marriage)"* is another common statement.

The third general factor that I believe influences becoming a therapist is the professional's willingness to use life experiences as part of his or her professional maturation. Once again the issue seems self-evident, and it is reasonable to ask if all therapists don't use their life experiences in the service of their work. For me the answer is "No," not all therapists use their experiences—at least not in the same way that some therapists do.

I wrote about this issue in describing some of what I learned about myself from spending many hours "being with" dying friends (Lewis, 1982). I suggested that those experiences brought me in touch with anxiety about my death and, in particular, with the strident defenses against dealing openly with that anxiety. These processes, I thought, made me somehow more available to patients struggling to deal with a variety of related concerns. Experiences with dying friends are, of course, unusual experiences, as are examples used in earlier chapters. Although more usual personal experiences are an important influence on one's work, I have found it is the more unusual experiences—dying friends and responses to near fatal automobile accidents—that have the power to break through one's customary defenses and lead to learning.

At the heart of this process of the ways in which personal experiences can facilitate development as a therapist is something of the willingness to face vulnerability. Becker (1973) said that no one would get out of bed without some underlying sense of omnipotence, some notion that come what may one will survive. I agree with this position, but am suggesting here a restricted paradoxical variant. To the extent that the psychotherapist deals openly with his or her vulnerability as experienced through life events, he or she is more apt to mature as a therapist.

Finally, the job is never done. One of the most exciting aspects of being a psychotherapist is that one continues to learn and grow. Mistakes may be the most common teacher, but these general factors certainly play a facilitating role.

CHAPTER 10

Summing Up

The argument underlying this book starts with the issues of complex and linear causality. I believe that both clinical and research evidence strongly suggests that psychopathology, including major psychiatric syndromes, is the result of complex interactions involving biological, psychological, and sociological variables. If we have learned anything in recent decades it is that describing a syndrome as solely brain disease, psychological conflict, or family dysfunction invariably oversimplifies. Always, multiple factors are involved, and the same syndrome in different individuals often reflects different combinations of etiologic variables.

In Chapter 1, factors that make the use of complex models of psychopathology like the biopsychosocial model difficult were described. I suggested that a true integration of such diverse variables is not yet possible and that the best that can be done currently is to use multiple perspectives to understand a patient's dilemma. Biological, psychological, and sociological perspectives are not inherently conflicting; rather, they can be understood as complementary views, much as one gets from viewing any entity or process from different vantage points.

If one gives up the hope of finding simple causal relationships in chronic processes ranging from arteriosclerosis to schizophrenia, if one accepts the inevitable uncertainties involved in having to rely on complex models, it is apparent that the psychiatrist must master the delicate art of exploration. I emphasize the connection between complex causality and exploration because in complex situations a method is needed to discover the linkages between multiple factors that in concert affect the probability of certain outcomes. Direct probes into complex fields rarely reveal linkages; their focus is too narrowly defined.

The exploration is based upon the professional's expertise in facilitating the patient's narrative. This facilitation is the heart and soul of the psychotherapeutic method. Without its mastery, the psychiatrist is hemmed in by the narrowness of structured interviews and manuals for the conduct of psychotherapy. However valuable such instruments are in the world of research quantification and descriptive diagnosis, major reliance upon them in the real world of clinical care result in a pallid charade.

The importance of facilitating the patient's narrative leads inevitably to recent interest in narrative theory. Are the narratives upon which the professional arrives at an understanding of the patient's dilemma historical truth or narrative truth (Spence, 1982)? To what extent does one's present circumstance modify his or her story of the past? To what extent do one's expectations of the future modify recall of the past? How influenced is one by cultural prescriptions of the "good" narrative? Although the patient's life experiences may modify the relationship with the doctor, how does the relationship with the doctor modify the patient's life narrative? Is there a relationship between certain types of doctor-patient relationship structures and the recall of particular stories from the past?

These issues lead to the dialectic about the nature of reality. Is there a real, "out there," discoverable reality existing independently from the observer's perceptions, or are different realities inevitably constructed by individuals with different perspectives?

If the need for exploratory skills arises from complex causality, what are the implications for the training of psychotherapists? I have emphasized the need to intensify our efforts to train systematically succeeding generations of psychiatrists in the crucial processes of psychotherapy. In sharing the evolution of my ideas and my experiences with many residents and other professionals, I have emphasized the need to experiment with methods of teaching. The seminar has been divided into phases or teaching modules, although such division is mostly a heuristic device. Facilitating exploration, sensitivity to affect and empathy, relationship structure, cognitive interventions, and a clinical formulation that includes an array of variables are major strands in the tapestry of the seminar. As such, they are interwoven; all are necessary for the whole.

My view of the current status of the real world of psychiatric training suggests much uncertainty in what will evolve. A discipline's core distinctions, those perspectives and processes that distinguish it from other disciplines, are strongly molded by what the members of a discipline actually do. If powerful economic and political forces increasingly mandate only the briefest of treatments, the focus of clinicians will come increasingly to be symptom-focused, descriptive, and pharmacological. If enough clinicians are forced into that pattern of clinical activity for long enough periods of time, psychiatry's definition of itself will gradually change, and symptoms, descriptions, and prescribing drugs will become our core distinctions. Patients with complex psychopathology, whether acute or chronic, will receive inadequate care.

From the inside, psychiatry is also threatened by the reductionism of the neurosciences. The amazing advances in understanding the biology of the brain are quickly converted by some into the call for a biological model of psychopathology, a model that either dismisses or gives but token attention to the role of psychological and social system variables.

A recent example of such reductionism is Cummings's (1990) announcement of a paradigm shift. He states that neurologic interpretations of behavior have replaced psychological explanations—neurobiological causes are now the "dominant explanatory theses." Idiopathic psychiatric illnesses are the product of brain dysfunction. Major psychiatric disorders, as well as more subtle behavioral traits, are genetically determined manifestations of brain function.

Although human suffering must be recognized and relieved, the emerging neuropsychiatric paradigm, according to Cummings, provides a comprehensive framework for understanding behavior, new and successful treatment strategies, compatibility with other scientific disciplines, the promise of better diagnostic and prognostic information, and a method for distinguishing psychiatry from other disciplines. Cummings calls for the adoption of the new neuropsychiatric paradigm in our training programs to replace the "smorgasbord of offerings" from many schools of thought. The promised land is available, and its terrain is biological reductionism.

Compare Cummings's view of psychiatry's future with recent vintage Havens (1989). Havens states that the future belongs to psychiatry because it is the only systematic study that takes as its central concern the biological, neurological, psychological, and sociological perspectives of the individual. The strength of psychiatry and its bright future are based on its capacity to draw on all these perspectives. It is through this comprehensiveness that the psychiatrist may lead and, in that leading, may delegate specific tasks to members of other disciplines.

Havens believes that what is meant by civilization is the replacement of cannibalism and physical enslavement by their psychological equivalents, psychological invasiveness and compliance. Psychiatry's claim to the future is buttressed by its role in dealing with these equivalents both by defining the conditions of human self-respect and freedom and by helping individuals achieve them.

These two views of the future of psychiatry are very different and call for remarkably different training programs. Cummings would reorient psychiatric residency training to a single paradigm, neuropsychiatry. Havens would continue to develop a comprehensive and pluralistic model of training. Psychotherapy training would be an important aspect of programs based on Havens's vision of the future, but of little significance in Cummings's prescription. Which view will win?

As I write these words, my attention is directed to the special article in *The American Journal of Psychiatry* (Mohl et al., 1990). It is a report from a committee representing two prestigious organizations centrally concerned with psychiatric education. Its title is precisely to the point: *Psychotherapy Training for the Psychiatrist of the Future*. The authors note that we are at a historic watershed in regard to the future of teaching psychodynamic psychotherapy. They suggest that although most programs retain a place for such training, the place is far smaller than before.

The committee focused its attention on long-term psychodynamic psychotherapy, that which has been variously called "expressive," "exploratory," "reconstructive," "psychodymanic," "psychoanalytic," or "insight-oriented" psychotherapy. They asked whether training in this type of psychotherapy should continue to be a significant part of a psychiatric residency and, if so, what a modern curriculum should contain.

The committee identified 10 major reasons to continue to teach psychodynamic psychotherapy. These are:

1. It is an effective treatment for many disorders.

2. It provides trainees with a unique opportunity for professional growth and development.
3. It provides a basis for the supervision of other professionals.
4. It provides a basis for more effective management of nonpsychotherapy doctor-patient relationships.
5. It enhances learning about other dyadic relationships.
6. It enhances interviewing techniques by providing the opportunity to observe longitudinally the processes present in an initial interview.
7. It introduces the trainee to an in-depth and longitudinal study of both conscious and unconscious mental functioning.
8. It provides access to material essential for understanding psychodynamics.
9. It provides the opportunity to anticipate, analyze, and avoid ethical dilemmas.
10. It facilitates the development of the skills of observation, hypothesis-formation, and data analysis.

Thus, the committee concludes that the exciting developments in neuroscience "do not displace the importance of the experiential meanings that behavioral and mental phenomena have to patients" (p. 9). They answer their first question affirmatively: Teaching psychodynamic psychotherapy should be an important part of residency training.

The committee then lays out a model curriculum. I was pleased that they cited my ideas as influential in the development of this curriculum. In particular, they noted the usefulness of a structured introduction to psychotherapy skills. The model curriculum is a rich, yet concise, articulation of training experiences at each of the four postgraduate years. The adoption of such a curriculum would do much to insure that the ability to facilitate exploration would continue to be one of psychiatry's core distinctions.

Unfortunately, in my view, the committee also spells out another curriculum for "minimal training in psychodynamic psychotherapy." Believing that some programs have "limited resources" to devote to psychotherapy training, they suggest that 200 hours of psychotherapy over the four-year residency involving at least four patients, a total of 100 hours of supervision, and one year of weekly didactic sessions constitute an acceptable minimum format. The committee believed that "some knowledge of the unconscious, transference, and therapeutic process is more desirable than ignorance or misunderstanding" (p. 12).

I disagree with this aspect of the committee's otherwise thoughtful report. The adage, "A little knowledge is a dangerous thing," applies directly to this recommendation. How, for example, would patients know that a particular "fully trained" psychiatrist was not competent to provide psychotherapy? Would additional certification of special competency be required? If so, how many certifications? Group psychotherapy, marital and family therapies, psychopharmacology, behavior therapies—the list could be long.

More to the point, however, is how would such a "minimally trained" psychotherapist know when his or her competence was adequate for a particular patient? How would such judgments about one's self be influenced by the economics of practice? I recall, for example, a group of older neuropsychiatrists who were in practice when I started my practice. These were men with primary training in neurology or neuropathology and, most important, very little, if any, training in psychodynamic psychotherapy. There were too few neurological patients in the community to keep these physicians busy, and many of them filled their days with psychotherapy patients. I believe many were not competent as psychotherapists. During the late stages of their careers, the issue of hospital privileges became sticky. Should such physicians be privileged to care for patients on open-staff general hospital units if that care appeared to involve psychotherapy was a question often debated.

I would not want to be cared for by any medical specialist who had a minimal curriculum in one of the skills generally assumed to be part of a particular specialist's repertoire. How would I know, for example, whether my cardiologist had a "model" or "minimal" curriculum in coronary arteriography, if that is what I needed? To be sure, I know how to get such information. The average patient, however, does not have ready access to such information and is often at the mercy of what the physician communicates.

Most physicians do not deliberately distort when it comes to areas of competence. They acknowledge what they can and cannot do and make use of ready referral to others with the needed competence. I am not concerned about a deliberately fraudulent misrepresentation. No, it is the human way in which we are all capable of deluding ourselves about specific areas of competence that concerns me. My neuropsychiatrist colleagues of yesteryear were not frauds. In certain economic contexts, however, they came to believe they had a level of psychotherapeutic competence that was not justified by their training or experience.

There is yet another issue. When I published my earlier book on psychotherapy training, I called it *To Be a Therapist: The Teaching and Learning* (1978). As I suggested in the preceding chapter, the verb form "to be" was used deliberately. I believe one becomes a therapist. The work of psychotherapy comes to mold one's experience of one's self. "Doing" psychotherapy is fundamentally different from being a therapist. Being a therapist, among other things, involves developing an ability to monitor interpersonal distance, to know where one is in regard to the experience of others. It involves a level of self-awareness that facilitates an acceptance of one's limitations. It includes the development of respect for boundaries, one's own as well as those of others. As such, the psychotherapist comes to respect a subjective reality; there are, in my opinion, few hard-liners in the world of psychotherapy.

I believe that other changes occur over time in the self-systems of psychotherapists. This is a little studied area, however, and there are no longitudinal data with which to confirm my impressions. A suggested perspective from which to study the longitudinal development of psychotherapists is that of ideology. Is there anything regular in the longitudinal development of the ideologies that direct the work of psychotherapists? Havens (1973) suggested that there are four major psychiatric ideologies or schools. The oldest is the objective-descriptive, and out of this school psychoanalysis developed as the second major ideology. From psychoanalysis come two new ideologies: interpersonal or social psychiatry and existential psychiatry.

It was Havens who suggested that the individual psychiatrist's development could be understood as passing through four stages, each of which corresponds to one of the schools. Thus, the young psychiatrist starts with an objective-descriptive orientation in which the accurate diagnosis of psychiatric disturbances (understood as "real" disease) is paramount. DSM-III-R is the absolute delight of the objective-descriptive psychiatrist for many diagnostic entities carry specific, usually pharmacologic, treatment recommendations.

For my generation of psychiatrists, the objective-descriptive ideology was, for all but a very few, a brief ideological sojourn before moving on to a psychoanalytic orientation. I suspect that in this day and age a greater proportion of young psychiatrists remain in the objective-descriptive stage, depending to a considerable extent on the nature of their training experiences.

Because for my generation the psychoanalytic approach was so compelling, many psychiatrists remained comfortably within that school for all or most of their careers. For others, however, the primary theoretical orientation became increasingly interpersonal. Most of the first generation leaders in the family therapy movement, for example, were both psychiatrists and psychoanalysts. For many of them, to adopt an interpersonal paradigm involved turning away from the earlier psychoanalytic orientation. Finally, some psychiatrists come later in their careers to adopt an existential ideology with its emphasis on ultimate concerns and the radical empathic "being" and "staying" with patients' deepest feelings.

Ralph (1980) collected data from trainees and their supervisors that support the idea of a developmental progression in the ideologies of psychotherapists.

I use these observations to support my belief that there are changes in the self-systems of psychotherapists over time, that those who only "do" therapy have somehow missed something very central to being a therapist. Perhaps the distinction is between career and calling. Although the idea of career involves greater investment of self than does working in order to make a living, a calling both carries more of the individual's identity and incorporates the idea of the common good.

In all of this I suggest that the beginning is crucial. How trainees are introduced to the processes of psychotherapy may influence all of their subsequent professional development. The model curriculum proposed by the committee is an excellent beginning. The minimal curriculum is not, and I for one would rather have those trainees whose primary interest is in neuroscience be exposed to the model curriculum or none at all. For me both the "none at all" option and the minimal curriculum mean that something central, a core distinction, has not been dealt with in the educational process.

Thus, decisions about curriculum ultimately reflect value judgments about psychiatry's core distinctions. When there is a rapid increase in knowledge within a particular domain, as there has been in the neurosciences in recent decades, time must be found in the curriculum for those advances. At such times we are apt to discard important insights from earlier eras. Or, as I believe is the case currently, the advances in neuroscience obscure exciting theoretical and research developments in other areas. I mention here only one such area, developmental psychology. I have become aware of this exciting, rapidly moving area as the result of participating in a longitudinal family research project in which marital and family system data are collected concurrently with infant and child developmental data (Lewis, 1989). The introduction by developmental psychologists of a variety of new approaches to the study of developmental experiences has remarkable implications for understanding psychopathology. With few exceptions, however, these studies are reported in developmental psychology journals and are not routinely seen by clinicians. If, for example, the excitement of attachment research is discussed with clinical colleagues, many are relatively unaware of these developments. It is as if the exciting clinical studies growing out of advances in neuroscience crowd out of view major advances in other areas.

Thus, there are a number of reasons for uncertainty and doubt about the future of psychotherapy training in psychiatry. I believe, however, that a future can be assured if those of us who believe strongly in its importance come together around a small number of constructs. At the minimum these include a continuing emphasis on and dialectic about psychiatry's core distinctions. I suggest that, in addition to the distinction of medicine's traditions and standards regarding patient care, psychiatry is characterized by its comprehensiveness. The characteristic of thinking broadly about the patient, of viewing the patient's dilemma from multiple perspectives, is based upon our recognition of the role of complex causality in psychopathology. From the importance of comprehensiveness comes the third core distinction: the reliance upon psychotherapeutic competence with its crucial exploratory base as an indispensable feature of our speciality.

Although we share physicianship with all medical specialties, and we share the commitment to comprehensiveness with several, it is our psychotherapeutic function with its emphasis on the patient's experience that marks our boundaries. Understanding the patient's experience from biological, psychological, and sociological perspectives insures consideration of a variety of treatment interventions. If this distinction erodes or is allowed to die, we can lose our identity.

Those who construct curricula are in a central role. We need to reassess the ways in which we introduce our students to psychotherapy. In that reassessment we need to acknowledge that truly understanding dyadic relationships generally—and the psychotherapeutic alliance specifically—is in its infancy. Exciting developments point the way to new methods of teaching the essential processes of psychotherapy.

As I look back at the evolution of the seminar over the 25-year period, several issues need emphasis. First, my teaching of psychotherapy continues to have an experimental quality. New approaches often come from the work of others, but sometimes they seem to arise from my own musings, which almost always concern the concrete details of constructing exercises that introduce students to important psychotherapeutic processes. Although new ideas and constructs arising from advances in theory are introduced to the students (often by giving them reprints), I have continued to emphasize doing—the experiencing of what therapists actually do in the course of psychotherapy.

I would also emphasize that my hope is to introduce beginning students to the sense of adventure that is central to psychotherapy. This task is better accomplished during the latter stages of the seminar; early on the students are simply too anxious and often feel lost in the terrain of an entirely new way of interviewing and being "with" another person. Once the basic skills of facilitating an exploration become internalized, students begin to relax and wonder with pleasure about where a particular interview will end. It is this sense of both wonder and pleasure that enhance enthusiasm and the development of a psychotherapist's adventurousness.

At a personal level the seminar has provided me with an ongoing opportunity for an intense involvement with bright and talented young people. For a teacher clearly in the late stage of his career, there are many advantages of such an involvement—but most of all I treasure the deep satisfaction of sharing what I have learned as a psychotherapist and the pleasure of doing something that feels useful. My experience is that doing so casts a glow on this stage of my life. This is a debt I owe to students past and present.

Appendix A

Candidate	______________________	Applicant should be offered a position
Applying for	______________________	
Date Interviewed	______________________	______ Yes ______ No

Rating Scale for Applicants for Residency Program

We are interested in obtaining quantifiable judgments regarding residency candidates. We ask that you use as a framework your involvement with recent classes of residents.

In rating these items, please proceed as follows: Think first if the individual performance is unacceptable, adequate, or superior. If so, rate 1, 3, or 5 respectively. If performance is not adequate but not bad enough to be unacceptable, rate 2. Similarly, if performance is somewhat better than adequate but not clearly superior, rate 4. Please try to avoid the "halo effect," i.e., tending to be influenced in rating a particular item by your rating on other items or by your overall impression of the individual.

	Circle the number which most closely reflects your judgment in the area described.
1. Human qualities considered to be significant in the helping professions (here rate evidence of warmth, genuineness, lack of rigidity of defenses).	1 2 3 4 5
2. Apparent intellectual skills (here rate verbal skills and your opinion of overall intellectual functions).	1 2 3 4 5

3. Motivation for psychiatry (here rate evidence which suggests the desire to enter the field is not based solely on personal problems). 1 2 3 4 5

4. General level of psychological maturity (here note both the absence of signs of immaturity, such as unusual dependency, passivity, demandingness, hostility, etc., and the presence or suggestion of age appropriate independence, ability to care for others, etc.). 1 2 3 4 5

5. General acceptability as a resident (here rate your personal interest in having this candidate as a member of our resident group. A rating of 5 would signify your enthusiastic wish to accept this candidate and a rating of 1 would indicate that you feel this candidate should not be accepted). 1 2 3 4 5

Additional comments (use back if needed):

____

Signature

Appendix B

BEAVERS-TIMBERLAWN FAMILY EVALUATION SCALE

Family Name .... Rater ....

Segment .... Date ....

Instructions: The following scales were designed to assess the family functioning on continua representing interactional aspects of being a family. Therefore, it is important that you consider the entire range of each scale when you make your ratings. Please try to *respond on the basis of the videotape data alone,* scoring according to what you see and hear, rather than what you imagine might occur elsewhere.

I. *Structure of the Family*

A. Overt Power: Based on the entire tape, check the term that best describes your general impression of the overt power relationships of this family.

1	1.5	2	2.5	3	3.5	4	4.5	5
Chaos		Marked dominance		Moderate dominance		Led		Egalitarian
Leaderless; no one has enough power to structure the interaction.		Control is close to absolute. No negotiation; dominance and submission are the rule.		Control is close to absolute, Some negotiation, but dominance and submission are the rule.		Tendency toward dominance and submission, but most of the interaction is through respectful negotiation.		Leadership is shared between parents, changing with the nature of the interaction.

B. Parental Coalitions: Check the terms that best describe the relationship structure in this family.

1	1.5	2	2.5	3	3.5	4	4.5	5
Parent-child coalition				Weak parental coalition				Strong parental coalition

C. Closeness

1	1.5	2	2.5	3	3.5	4	4.5	5
Amorphous, vague and indistinct boundaries among members				Isolation, distancing				Closeness, with distinct boundaries among members

II. *Mythology:* Every family has a mythology; that is, a concept of how it functions as a group. Rate the degree to which this family's mythology seems congruent with reality.

1	1.5	2	2.5	3	3.5	4	4.5	5
Very congruent		Mostly congruent				Somewhat incongruent		Very incongruent

III. *Goal-Directed Negotiation:* Rate this family's overall efficiency in negotiating problem solutions.

1	1.5	2	2.5	3	3.5	4	4.5	5
Extremely efficient		Good				Poor		Extremely inefficient

IV. *Autonomy*

A. Clarity of Expression: Rate this family as to the clarity of disclosure of feelings and thoughts. This is not a rating of the intensity or variety of feelings, but rather of clarity of individual thoughts and feelings.

1	1.5	2	2.5	3	3.5	4	4.5	5
Very clear				Somewhat vague and hidden				Hardly anyone is ever clear

B. Responsibility: Rate the degree to which the family members take responsibility for their own past, present, and future actions.

1	1.5	2	2.5	3	3.5	4	4.5	5
Members regularly are able to voice responsibility for individual actions				Members sometimes voice responsibility for individual actions, but tactics also include sometimes blaming others, speaking in 3rd person or plural				Members rarely, if ever, voice responsibility for individual actions

C. Invasiveness: Rate the degree to which the members speak for one another, or make "mind reading" statements.

1	1.5	2	2.5	3	3.5	4	4.5	5
Many invasions				Occasional invasions				No evidence of invasions

D. Permeability: Rate the degree to which members are open, receptive and permeable to the statements of other family members.

1	1.5	2	2.5	3	3.5	4	4.5	5
Very open		Moderately open				Members frequently unreceptive		Members unreceptive

V. *Family Affect*

A. Range of Feelings: Rate the degree to which this family system is characterized by a wide range expression of feelings.

1	1.5	2	2.5	3	3.5	4	4.5	5
Direct expression of a wide range of feelings		Direct expression of many feelings despite some difficulty		Obvious restriction in the expressions of some feelings		Although some feelings are expressed, there is masking of most feelings		Little or no expression of feelings

B. Mood and Tone: Rate the feeling tone of this family's interaction.

1	1.5	2	2.5	3	3.5	4	4.5	5
Usually warm, affectionate, humorous and optimistic		Polite, without impressive warmth or affection; or frequently hostile with times of pleasure		Overtly hostile		Depressed		Cynical, hopeless and pessimistic

C. Unresolvable Conflict: Rate the degree of seemingly unresolvable conflict.

1	1.5	2	2.5	3	3.5	4	4.5	5
Severe conflict, with severe impairment of group functioning		Definite conflict, with moderate impairment of group functioning		Definite conflict, with slight impairment of group functioning		Some evidence of unresolvable conflict, without impairment of group functioning		Little, or no unresolvable conflict

D. Empathy: Rate the degree of sensitivity to, and understanding of, each other's feelings within this family.

1	1.5	2	2.5	3	3.5	4	4.5	5
Consistent empathic responsiveness		For the most part, an empathic responsiveness with one another, despite obvious resistance		Attempted empathic involvement, but failed to maintain it		Absence of any empathic responsiveness		Grossly inappropriate responses to feelings

VI. *Global Health-Pathology Scale: Circle the number* of the point on the following scale that best describes this family's health or pathology.

10	9	8	7	6	5	4	3	2	1
Most Pathological									Healthiest

References

Ainsworth, M., Blehar, M, Waters, E., & Wall, S. (1978). *Patterns of Attachment.* Hillside, NJ: Lawrence Erlbaum Associates.

Balint, M. (1972). *The Doctor, His Patient, and His Illness.* New York: International Universities Press.

Bateson, G. (1972). *Steps to an Ecology of the Mind.* New York: Ballantine Books.

Beahrs, J. O. (1986). *Limits of Scientific Psychiatry.* New York: Brunner/Mazel.

Beavers, W. R. (1977). *Psychotherapy and Growth: A Family Systems Perspective.* New York: Brunner/Mazel.

Becker, E. (1973). *The Denial of Death.* New York: The Free Press.

Bowlby, J. (1973). *Attachment and Loss: Vol. 2. Separation.* New York: Basic Books.

Bowlby, J. (1980). *Attachment and Loss: Vol. 3. Loss, Sadness, and Depression.* New York: Basic Books.

Bowlby, J. (1982). *Attachment and Loss: Vol. 1. Attachment* (2nd ed.). New York: Basic Books.

Butler, S. F., & Binder, J. L. (1987, August). Cyclical psychodynamics and the triangle of insight: An integration. *Psychiatry, 50,* 218–231.

Churchill, L. R. (1989, July). Ethical issues in chronic, severe, and catastrophic illnesses. *Southern Medical Journal, 82*(7), 815–817.

Coles, R. (1989). *The Call of Stories.* Boston: Houghton Mifflin, Co.

Cummings, J. L. (1990, January). Neuropsychiatry: The paradigm shift. *The Psychiatric Times.*

Dell, P. F. (1989). Violence and the systemic view: The problem of power. *Family Process, 28,* 1–14.

Eisenberg, L. (1988). Quoted in Barnes, D. M. Psychiatrists psych out the future. *Science, 242,* 1013–1014.

Engel, G. L. (1980). The clinical application of the biopsychosocial model. *The American Journal of Psychiatry, 137*(5), 535–544.

Evelyn, R. (1989, July). Psychiatrists would be hurt by proposed medicare payment schedule. *The Psychiatric Times.*

Fabrega, H., & Manning, K. (1973). An integrated theory of disease: Lodino-Mestizo views of disease in the Chiapas Highlands. *Psychosomatic Medicine, 35*(3), 223–299.

Fowles, J. (1983). *The Tree.* New York: The Ecco Press.

Freudenheim, M. (1989, June 29). Debating Canadian health model. *The New York Times.*

Friedman, R. S., & Lister, P. (1987). The current status of psychodynamic formulation. *Psychiatry, 50,* 126–141.

Fuchs, V. R. (1974). *Who Shall Live? Health, Economics, and Social Choice.* New York: Basic Books.

Gay, P. (1988). *Freud: A Life for Our Time.* New York: W. W. Norton.

Gossett, J. T., Lewis, J. M., & Barnhart, F. D. (1983). *To Find a Way: The Outcome of Hospital Treatment of Disturbed Adolescents.* New York: Brunner/Mazel.

Greenberg, L. S., & Safran, J. D. (1989). Emotion in psychotherapy. *American Psychologist, 44*(1), 19–29.

Greenson, R. R. (1960). Empathy and its vicissitudes. *International Journal of Psychoanalysis, 41,* 418–424.

Gross, J. (1989, March 27). What medical care the poor can have: Lists are drawn up. *The New York Times.*

Haley, J. (1980). *Leaving Home: The Therapy of Disturbed Young People.* New York: McGraw-Hill.

Halleck, S. (1974). *Politics of Therapy.* New York: Aronson.

Harrison, T., Beeson, P. (1950). *Principles of Internal Medicine,* (1st ed.). Philadelphia: The Blakiston Company.

Havens, L. L. (1973). *Approaches to the Mind: Movement of the Psychiatric Schools from Sects toward Science* Boston: Little, Brown, & Co.

Havens, L. L. (1986). *Making Contact.* Cambridge: Harvard University Press.

Havens, L. L. (1989). *A Safe Place: Laying the Groundwork of Psychotherapy.* Cambridge, MA: Harvard University Press.

Hazan, C., & Shaver, P. (1987). Romantic love conceptualized as an attachment process. *Journal of Personality and Social Psychology, 52,* 511–524.

Johnson, J. (1989, July 8). Study finds 12 percent of American young have mental disorders. *The New York Times.*

Kagan, N. (1967, December). *Studies in Human Interaction,* 3 vols. U. S., ED107946. Dept. H. E. W.

Kantor, D., & Lehr, W. (1975). *Inside the Family: Toward a Theory of Family Process.* San Francisco: Jossey-Bass.

Kramer, P. D. (1989). *Moments of Engagement: Intimate Psychotherapy in a Technological Age.* New York: W. W. Norton & Company.

Lamm, R. D. (1989). Critical decisions in medical care: Birth to death. *Southern Medical Journal, 82*(7), 822–824.

Laughlin, H. P. (1967). *The Neuroses.* Washington: Butterworth.

Levinson, D. (1978). *The Seasons of a Man's Life.* New York: Ballantine Books.

Lewis, J. M. (1978). *To Be a Therapist: The Teaching and Learning.* New York: Brunner/Mazel.

Lewis, J. M. (1979). *How's Your Family?* New York: Brunner/Mazel.

Lewis, J. M. (1979a). The inward eye: Monitoring the process of psychotherapy. *Journal of Continuing Education in Psychiatry,* 17–26.

Lewis, J. M. (1982). Dying with friends: Implications for the psychotherapist. *American Journal of Psychiatry,* 261–266.

Lewis, J. M. (1988). The transition to parenthood: I. The rating of prenatal marital competence. *Family Process, 27,* 149–165.

Lewis, J. M. (1988a). The transition to parenthood: II. Stability and change in marital structure. *Family Process, 27*, 273–283.

Lewis, J. M. (1989). *The Birth of the Family: An Empirical Inquiry*. New York: Brunner/Mazel.

Lewis, J. M., Beavers, W. R., Gossett, J. T., & Phillips, V. A. (1976). *No Single Thread: Psychological Health in Family Systems*. New York: Brunner/Mazel.

Lewis, J. M., & Looney, J. G. (1983). *The Long Struggle: Well-functioning, Working-class Black Families*. New York: Brunner/Mazel.

Lewis, J. M., Owen, M. T., & Cox, M. J. (1988). The transition to parenthood: III. Incorporation of the child into the family. *Family Process, 27*, 411–421.

Luborsky, L. (1976). Helping alliances in psychotherapy. In J. L. Claghorn (Ed.), *Successful Psychotherapy*. New York: Brunner/Mazel, 92–111.

Luborsky, L. (1977). Measuring a pervasive psychic structure in psychotherapy: The Core Conflictual Relationship Theme. In N. Freedman & S. Grand (Eds.), *Communicative Structures and Psychic Structures*. New York: Plenum Press, 367–395.

Luborsky, L., McLellan, A. T., Woody, G. E., O'Brien, C. P., & Auerbach, A. (1985). Therapist success and its determinants. *Archives of General Psychiatry, 42*, 602–611.

Luborsky, L., Crits-Christoph, P., Mintz, J., & Auerbach, A. (1989). *Who Will Benefit from Psychotherapy: Predicting Therapeutic Outcomes*. New York: Basic Books.

Melges, F. T., & Swartz, M. S. (1989). Oscillations of attachment in borderline personality disorder. *Journal of the American Psychiatric Association, 146*(9), 115–120.

Mohl, P. C., Lomax, J., Tasman, A., Chan, C., Sledge, W., Summergrad, P., & Notman, M. (1990). Psychotherapy training for the psychiatrist of the future. *American Journal of Psychiatry, 147*(1), 7–13.

Myers, J. M. (1984). *Cures by Psychotherapy: What Effects Change?* New York: Praeger.

Owen, M. T., & Henderson, V. D. (1989, April). Relations between child care qualities and child behavior at age 4: Do parent-child interactions play a role? Paper presented at the biennial meetings of the Society for Research in Child Development. Kansas City.

Perry, S., Cooper, A. M., & Michels, R. (1987). The psychodynamic formulation: Its purpose, structure, and clinical application. *American Journal of Psychiatry, 144*(5), 543–550.

Ralph, N. B. (1980). Learning psychotherapy: A developmental perspective. *Psychiatry, 43*, 243–250.

Robins, L. N. (1966). *Deviant Children Grown Up*. Baltimore: The Willams & Wilkins Company.

Rogers, C. R. (1961). *On Becoming a Person*. Boston: Houghton Mifflin.

Rokeach, M. (1973). *The Nature of Human Values*. New York: The Free Press.

Rutter, M. (1982). Protective factors in children's responses to stress and disadvantage. In M. W. Kent & J. E. Rolf (Eds.), *Primary Prevention of Psychopathology. Vol. 3. Promoting Social Competence and Coping in Children*. Hanover, NH: University Press of New England, 49–74.

Rutter, M., & Quinton, D. (1984). Long-term follow-up of women institutionalized in childhood: Factors promoting good functioning in adult life. *British Journal of Development Psychology, 2*, 191, 204.

Safire, W. (1989, July 30). Who's in the details? *The New York Times Magazine*, 8 & 10.

Seligman, M. E. P. (1989). Explanatory style: predicting depression achievement, and health. In M. D. Yapko (Ed.), *Brief Therapy Approaches to Treating Anxiety and Depression*. New York: Brunner/Mazel, 5–32.

Spence, D. P. (1982). *Narrative Truth and Historical Truth: Meaning and Interpretation in Psychoanalysis*. New York: W. W. Norton & Company.

Spiegel, J. P., & Kluckhohn, F. R. (1971). Integration and conflict in family behavior. In J. P. Spiegel, *Transactions*. New York: Science House, 143–187.

Stern, D. N. (1985). *The Interpersonal World of the Infant*. New York: Basic Books.

Thomas, A., & Chess, S. (1977). *Temperament and Development*. New York: Brunner/Mazel.

Thomas, L. (1974). *The Lives of a Cell: Notes of a Biology Watcher*. New York: The Viking Press.

Tienari, P., Sorri, A., Lahti, I., Naarala, M., Wahlberg, K., Ronkko, T., Pohjola, J., & Moring, J. (1985). The Finnish adoptive family study of schizophrenia. *The Yale Journal of Biology & Medicine, 58*, 227–237.

Tomm, K. (1987). Interventive interviewing: Part I. Strategizing as a fourth guideline for the therapist. *Family Process, 26*, 3–13.

Tomm, K. (1987a). Interventive interviewing: Part II. Reflexive questions as a means to enable self-healing. *Family Process, 26*, 167–183.

Tomm, K. (1988). Interventive interviewing: Part III. Intending to ask lineal, circular, strategic, or reflexive questions? *Family Process, 27*, 1–15.

Traux, C. B., & Carkhuff, R. R. (1967). *Toward Effective Counseling and Psychotherapy*. Chicago: Aldine Publishing Company.

Vaillant, G. E. (1971). Theoretical hierarchy of adaptive ego mechanisms. *Archives of General Psychiatry, 24*, 107–118.

Whitney, C. R. (1989, June 26). Thatcher's new health plan: An outcry rises on all sides. *The New York Times*.

Witmer, H. L. (Ed.) (1947). *Teaching Psychotherapeutic Medicine: An Experimental Course for General Physicians*. New York: The Commonwealth Fund.

Suggested Reading for Students

Basch, M. F. (1988). *Understanding Psychotherapy.* New York: Basic Books.

Becker, E. (1973). *The Denial of Death.* New York: The Free Press.

Butler, S. F., & Binder, J. L. (Aug., 1987). Cyclical psychodynamics and the triangle of insight: An integration. *Psychiatry, 50,* 218–231.

Cassell, E. J. (1976). *The Healer's Art: A New Approach to the Doctor-Patient Relationship.* Philadelphia: Lippincott.

Celani, D. (1976). An Interpersonal approach to hysteria. *American Journal of Psychiatry, 133*(12), 1414–1418.

Coyne, J. C. (1976). Toward an interactional description of depression. *Psychiatry, 39,* 28–40.

Dell, P. F. (1989). Violence and the systemic view: The problem of power. *Family Process, 28,* 1–14.

Eisenberg, L. (April 14, 1972). The human nature of human nature. *Science, 176*(4031), 123–128.

Engel, G. L. (1980). The clinical application of the biopsychosocial model. *The American Journal of Psychiatry, 137*(5), 535–544.

Frank, J. D. (1973). *Persuasion and Healing.* Baltimore: The Johns Hopkins University Press.

Friedman, R. S., & Lister, P. (1987). The current status of psychodynamic formulation. *Psychiatry, 50,* 126–141.

Gossett, J. T., Lewis, J. M., & Barnhart, F. D. (1983). *To Find a Way: The Outcome of Hospital Treatment of Disturbed Adolescents.* New York: Brunner/Mazel.

Hare-Mustin, R. T., & Marecek, J. (1986). Autonomy and gender: Some questions for therapists. *Psychotherapy, 23*(2), 205–212.

Havens, L. L. (1973). *Approaches to the Mind.* Boston: Little, Brown, & Co.

Havens, L. L. (1978). Explorations in the uses of language in psychotherapy: Simple empathic statements. *Psychiatry, 41,* 336–345.

Havens, L. L. (1979). Exploration in the uses of language in psychotherapy: Complex empathic statements. *Psychiatry, 42,* 40–48.

Havens, L. L. (1986). *Making Contact.* Cambridge: Harvard University Press.

Havens, L. L. (1989). *A Safe Place: Laying the Groundwork of Psychotherapy.* Cambridge: Harvard University Press.

Kramer, P. D. (1989). *Moments of Engagement: Intimate Psychotherapy in a Technological age.* New York: W. W. Norton & Company.

Lewis, J. M. (1979). The inward eye: Monitoring the process of psychotherapy. *Journal of Continuing Education in Psychiatry*, 17–26.

Lewis, J. M. (1982). Dying with friends: Implications for the psychotherapist. *American Journal of Psychiatry*, 261–266.

Luborsky, L. (1977). Measuring a pervasive psychic structure in psychotherapy: The core conflictual relationship theme. In N. Freedman & S. Grand (Eds.). *Communicative Structures and Psychic Structures*. New York: Plenum Press, 367–395.

Luborsky, L., McLellan, A. T., Woody, G. E., O'Brien, C. P., & Auerbach, A. (1985). Therapist success and its determinants. *Archives of General Psychiatry, 42*, 602–611.

MacKinnon, R., & Michels, R. (1971). *The Psychiatric Interview in Clinical Practice*. Philadelphia: W. B. Saunders Company.

Perry, S., Cooper, A. M., & Michels, R. (1987). The psychodynamic formulation: Its purpose, structure, and clinical application. *American Journal of Psychiatry, 144*(5), 543–550.

Rokeach, M., & Ball-Rokeach, S. J. (1989). Stability and change in American value priorities. *American Psychologist*, 775–784.

Scheflen, A. E. (1965). Quasi-courtship behavior in psychotherapy. *Journal for the Study of Interpersonal Process, 28*, 245–257.

Seligman, M. E. P. (1989). Explanatory style: predicting depression achievement, and health. In M. D. Yapko (Ed), *Brief Therapy Approaches to Treating Anxiety and Depression*. New York: Brunner/Mazel, 5–32.

Stern, D. N. (1985). *The Interpersonal World of the Infant*. New York: Basic Books.

Vaillant, G. E. (1971). Theoretical hierarchy of adaptive ego mechanisms. *Archives of General Psychiatry, 24*, 107–118.

Yalom, I. D. (1980). *Existential Psychotherapy*. New York: Basic Books.

Index

Note: Italicized page numbers indicate illustrations.